General Preface to the Series

It is no longer possible for one textbook to cover the whole field of Biology and to remain sufficiently up to date. At the same time teachers and students at school, college or university need to keep abreast of recent trends and know where the most significant developments are taking place.

To meet the need for this progressive approach the Institute of Biology has for some years sponsored this series of booklets dealing with subjects specially selected by a panel of editors. The enthusiastic acceptance of the series by teachers and students at school, college and university shows the usefulness of the books in providing a clear and up-to-date coverage of topics, particularly in areas of research and changing views.

Among features of the series are the attention given to methods, the inclusion of a selected list of books for further reading and, wherever possible, suggestions for practical work.

Readers' comments will be welcomed by the author or the Education Officer of the Institute.

1977   The Institute of Biology
                                          41 Queens Gate,
                                          London, SW7 5HU

60p

Preface

Crops and weeds are two great classes of plants of which some members are familiar to everybody. Directly or indirectly crops provide our food and many of the necessities of life. Weeds almost always interfere with the growth of crops and are important because of the huge losses in yield which they may cause, but they are also fascinating because of the many ways in which they have become adapted to a way of life inextricably bound up with man and his activities. I have tried to explain in this book some of the reasons why weeds are successful.

The book is written from the viewpoint of a plant physiologist interested in the ways in which plants are organized and how they cope with their environment. I hope readers will learn enough from it to look with a new respect at weeds even whilst waging war on them.

I am grateful to all those who have kindly given permission for the reproduction of material from other publications. Special thanks are due to my wife Mary for her work on the illustrations, to John Marshall for the photographs and particularly to Mr R. J. Chancellor of the A.R.C. Weed Research Organization, Oxford, who kindly read and commented most helpfully on the whole manuscript, though I remain responsible for any errors or omissions. The book is dedicated to Professor I. W. Selman who, by persuading me to give some lectures on weeds, first introduced me professionally to this marvellous group of plants.

Wye, 1977   T. A. H.

Contents

1   Introduction

Although the word 'weed' means something quite clear to almost everyone it is not easy to give a wholly satisfactory definition of this class of plants. The simplest and most common definition is 'any plant growing where it is not wanted'. This statement contains one very important and central idea about weeds, which is that they are exclusively associated with man and his activities. The disadvantage of the definition is that it does not fully allow for the concept of what is generally called 'weediness'. When we look at examples of weeds it soon becomes clear that there are some types of plant which appear to make good weeds and others which do not. Thus we may point to a poppy plant growing somewhere where it is doing no harm and still want to say 'that is a weed'. This is really a shorthand way of saying 'that plant shows characteristics associated with weeds'. Weediness may be defined as that group of characteristics possessed by plants which tend to occupy habitats associated with weeds. Such habitats have often been created wholly or partly by man.

Before the advent of man there were certainly disturbed and open habitats caused by natural agencies such as floods, landslides, glaciation, earthquakes and other agencies. The plants which were able to colonize and thrive in such places were probably those which, with hindsight, we might designate as 'weedy'. In other words the quality of weediness antedates man's arrival on the scene whilst weeds, in the sense of the definition given above, do not.

A more explicit definition of a weed, and one which provides a basis for almost everything which is said about these plants in this book, is given by BAKER (1965). This allows for the concept of weediness without difficulty whilst still making clear the relationship of weeds to man. According to Baker a plant is a weed 'if, in any specified geographical area, its populations grow entirely or predominantly in situations markedly disturbed by man (without, of course, being deliberately cultivated plants)'.

Weed biology is a subject embracing many disciplines including ecology, physiology and genetics and nobody can be equally expert in all of these. Of the many people all over the world who devote their whole time to the study of weeds, most are concerned directly or indirectly with weed control. The control of weeds is naturally an extremely important subject in its own right, but in these pages it is not possible to discuss it except in passing. Many of the works cited as references, however, do contain introductions to the study of weed control. The particular purpose of this book is to introduce weeds mainly as a biological phenomenon—a group of plants with a fascinating history and worthy of study for their own sakes. A good subtitle for the book might be 'an introduction to weediness and some of its consequences'.

Because there is no doubt about their importance to man and because nearly everybody is familiar with at least some examples it is in one way fairly easy to write about weeds. Amongst the problems of approaching such a large subject, however, is the sheer number of plants involved. Out of the whole world flora only a very tiny fraction, probably amounting to no more than a few hundred species, is composed of weeds. However, as soon as examples of their activities are discussed names multiply and tend to bewilder those not familiar with the plants. I have tried, therefore, to keep to an absolute minimum the number of plants referred to. In spite of this the list of plants named is quite long. Most of these are found in the British Isles, but a considerable number are much more widely distributed. I have used both Latin and common names in the text, and both are given on the first mention of any species. All the plants referred to are listed in an Appendix at the end of the book, in alphabetical order of Latin names. Common names, families and a note of whether the plant is annual, biennial or perennial are also included, though in the case of common names difficulties often arise because of national and regional variations. A fair proportion of the plants, and especially the common ones, are referred to many times and those who are not familiar with them by name may find it helpful to look at pictures of some of them. Those in KEBLE MARTIN (1965) are readily accessible, and useful pictures of many are also found in the book by HANF (1972). For the sake of uniformity the common names of all the British plants are taken from the book by DONY et al. (1974).

An interesting feature which emerges from the list of families contained in the Appendix is that some of these appear a disproportionate number of times. This is true in general of the world weed flora in which there is a predominance of plants from relatively highly advanced families. Table 1 shows the situation in relation to one list of 700 North American weeds.

Table 1   List of the seven families whose species make up 60% of the 700 species of weeds introduced into eastern North America, with the number of such species in each. (After FOGG (1975). Also cited by MUZIK (1970) from an earlier paper.)

Family	Number of species
Compositae	112
Gramineae	65
Cruciferae	62
Labiatae	60
Leguminosae	54
Caryophyllaceae	37
Scrophulariaceae	30

1.1   The importance of weeds

The scale of the problem of weeds, when considered on a worldwide basis, is enormous. The sophisticated agricultural methods employed in much of Europe, North America and in other developed countries tend to prevent our noticing the problem, especially since some of our more spectacular weeds such as the field poppies (*Papaver* spp.) and charlock (*Sinapis arvensis*) have been largely controlled. The problem is still very much with us, however, as the enormous annual bill for herbicides and the considerable crop losses due to uncontrolled weeds will testify. In areas without access to herbicide technology a very significant part of the physical process of cropping is still devoted to the relentless task of weed removal.

Weeds cause losses and inconvenience to man in many ways (see chapter 2) but the one to which attention is most often directed is loss of crop yield. Such losses are extremely difficult to estimate, especially since diseases due to fungi, bacteria and viruses also deplete yields, as do insects and other pests. A difference between these sources of loss is that weeds do most of the damage whilst the crop is growing, whilst many diseases and pests can also cause serious losses after harvest. In spite of the difficulties, estimates of crop losses have been made for a wide variety of crops, and CRAMER (1967) has collected an enormous volume of data on the subject. He compares losses due to pests, diseases and weeds for a whole range of crops both on a country by country basis and as world totals. A significant fact which appears from the data is the very wide range of losses which have been reported even for individual crops. Broadly

Table 2   Selected figures showing annual losses of crops to pests, diseases and weeds. (Data of CRAMER, 1967.) Figures are calculated on a world basis unless otherwise stated, and in the first three lines are in millions of tons.

Crop	Potential production in millions of tons	Losses due to Pests	Diseases	Weeds
All cereals	1467.5	203.7	135.3	167.4
Sugar beet and sugar cane	1330.4	228.4	232.3	175.1
Vegetable crops	279.9	23.4	31.1	23.7
All crops (% of potential value):				
(a) Worldwide	100	13.8	11.6	9.5
(b) Europe	100	5.1	13.1	6.8
(c) North & Central America	100	9.4	11.3	8.0
(d) Africa	100	13.0	12.9	15.7
(e) Asia	100	20.7	11.3	11.3

speaking losses due to pests, diseases and weeds are all of the same order, though again there are wide discrepancies. For example losses for fruit crops in Africa are given as 25% due to weeds, and 20% due to pests and diseases combined. In the case of potatoes, on the other hand, the figures are 6.5% due to pests, 21.8% due to diseases and only 4% due to weeds. It should be borne in mind that these three sources of loss may well affect each other. For example a diseased crop may be more susceptible to weed competition than a healthy one.

Table 2 shows some figures selected from Cramer's book, and even when due allowance is made for inaccuracies of estimation the picture which emerges is one which emphasizes the importance of weeds as a serious source of crop loss. Associated with loss of yield is, of course, an effect on price, which may be just as important from the point of view of the consumer, especially in a peasant economy.

1.2   The history of weeds

In his fascinating and readable book *Plants, Man and Life* ANDERSON (1954) comments that the history of weeds is the history of man. That this is bound to be true to some extent is implicit in the definitions given earlier, but if we look for example at the weed flora of Great Britain it is clear that many plants which are now regarded as weeds were present long before man came on the scene. Evidence for this comes from the study of plant remains and especially from the analysis of deposits in peat bogs. Peat is laid down steadily in relatively undisturbed layers, and pollen from surrounding vegetation is particularly well preserved in the conditions it provides. The major study of the history of the British flora published by GODWIN (1956) was based to a large extent on work of this kind and a good introduction to it is given by PENNINGTON (1974).

The post-glacial period in Britain extended from about 8300 B.C. At this time *Rumex* spp. (docks) and ribwort plantain (*Plantago lanceolata*) were both well established, and both are now regarded as characteristic weeds of open situations. As far back as the Middle Pleistocene period, about 600 000 years ago, the Cromer Forest Bed deposits show remains of the weedy plants chickweed (*Stellaria media*), knotgrass (*Polygonum aviculare*) and sheep's sorrel (*Rumex acetosella*). What is clear is that during the interglacial periods in the last glaciation, and also in the post-glacial period, there must have been many open and disturbed habitats which could well have been suitable for colonization by weedy species.

The evidence for a very long history for weeds in Britain is thus very strong, but it is interesting to consider what happened during the period from 8300 B.C. to the beginning of Neolithic cultivation in about 3000 B.C. The prevailing forest conditions were not suitable for the growth of weedy species, and yet these plants were apparently present in certain places and were thus able to colonize as soon as artificially disturbed sites became available again. There must always have been small local areas of disturbance due to natural causes such as rivers, but another likely possibility is that many plants of open habitats survived this period in the regions near sea shores or on higher mountain slopes where open conditions were maintained by the general physical environment. The advantage of such situations is that they are more or less permanent, whilst it is characteristic of other types of open habitat that the ecological succession proceeds relatively rapidly to give a closed community. Under modern conditions weeds and plants with weedy characteristics are frequently the pioneers of secondary successions caused by man-made or natural disturbances of the environment, but in many cases this weedy phase is quite brief.

When man first appeared on the scene in Britain he was a nomadic hunter, but even by this mode of life he undoubtedly affected the environment, albeit locally and transiently. Wherever there are men there is rubbish. ANDERSON (1954) points out that the rubbish heaps of primitive man were an ideal site for the establishment of weeds, and it seems extremely probable that it was from amongst the plants growing around his rubbish heaps that man began unconsciously to select his first crop plants. With the later development of the more settled life of the planter, rubbish heaps would tend also to be sites for the propagation and maintenance of the weedy species gathered with the crop and thrown away later. There is also ample evidence that many weed species were also used for food by early man, though this practice is by no means confined to the past. In many tropical areas weedy species are still eaten to supplement other crops.

Many of our present-day weeds thus have a long history in Britain, but a great many others were introduced from Europe much later by successive groups of colonizers. Examples of these are some of the weeds introduced by the Romans, which include the corn marigold (*Chrysanthemum segetum*), the red dead-nettle (*Lamium purpureum*) and the prickly sow-thistle (*Sonchus asper*). A very large number of weeds has been introduced accidentally by man, and in the countries of the New World which were not exposed to glaciation and were perhaps therefore less rich in the environments suitable for the development and evolution of the weedy habit, it is notable that an extremely large proportion of the present day weed flora is of alien plants. Examples of important and serious weeds in the United States which were introduced from Europe include perforate St John's wort (*Hypericum perforatum*), creeping thistle (*Cirsium arvense*) and field bindweed (*Convolvulus arvensis*).

We know a good deal about weeds within historical times, since they are referred to in books dealing with agriculture. SALISBURY (1961) quotes Fitzherbert's *Boke of Husbandry* of 1523 as commenting on the seriousness of thistles, nettles and charlock as weeds. In some cases weeds which were serious pests in earlier centuries remain so today, but in others changes in agricultural technology and practice have reduced previously serious weeds to relatively insignificant proportions. The corncockle (*Agrostemma githago*), formerly a very serious weed of cereals in England, is now relatively rare because its large seeds are easily removed from seed crops and populations rapidly fall unless replenished from outside because the seeds have only a rather short viable period in the soil. Many other examples could be quoted (see SALISBURY (1961), chapter 2). On the other hand some very common British weeds are of surprisingly recent origin. The winter wild oat (*Avena ludoviciana*) was introduced in 1917 and pineappleweed (*Matricaria matricarioides*) has only been with us for a little more than 100 years (SALISBURY, 1961, chapter 3; ROBERTS, CHANCELLOR and THURSTON, 1968).

Since man began to create disturbed environments on a large scale it is clear that enormous new possibilities have been opened up for weeds, and it is a striking fact that many weeds which are a serious problem in areas to which they have spread are relatively harmless in the places from which they were introduced. This matter is discussed further in chapter 5. Another phenomenon which has been recorded a number of times is that of a plant which has been known for a considerable time in one area and which relatively suddenly breaks out of that area as a weed. A particularly well documented case of this is that of the Oxford ragwort (*Senecio squalidus*) which was originally grown in the Oxford Botanic Garden around the end of the eighteenth century and was confined to Oxford for a considerable number of years but then spread rapidly to many parts of the country during the middle of the nineteenth century. Like the rosebay willowherb, which offers an equally spectacular story, it is a plant well adapted to sites where there has been burning; both are often found on derelict building sites and were very common on bombed sites during and after the second world war.

As a final note it is worth re-emphasizing the point mentioned earlier, that some weedy plants were certainly selected by primitive man as crops. Amongst crops thought to be have been selected and evolved from weedy ancestors are potatoes, carrots, sunflowers, barley, oats and rye; the weedy grass *Aegilops* is known to be an ancestor of modern wheat varieties. Thus weeds can be important to man in many ways, not all of them disadvantageous.

2 What Weeds Do

In the general sense in which we have been using the word 'weed' it is clear that what weeds do is to interfere in some way with man's use of land. The main thing which comes to mind is the interference by weeds with the growth of crops, which leads to an effect on yield, but there are other effects of weeds which are also important, though perhaps less obvious.

Before looking in more detail at some of the effects of weeds there is one question of terminology which it will be useful to mention. The word 'interference' in the previous paragraph was chosen deliberately for the purpose, though it is also very common for the word 'competition' to be used in relation to the effect of weeds on crop plants. There is a real difficulty here, because when a physiologist speaks of 'competition' in the biological sense there is an implicit understanding that what is meant is competition for *something*—some factor provided by the environment which is limited in some way such that if one organism obtains it another organism is thereby deprived of it. It may well be that many weed/crop situations are like this, and there are occasions when it seems appropriate to speak in these terms, but very often what one is faced with is a complex situation in which it is far from clear what is being competed for, and indeed whether 'competition' in the strict sense is taking place at all. With these and other difficulties in mind HARPER (1961) proposed that the more general term 'interference' should normally be used in relation to the effects of organisms on one another simply because it has a rather general meaning which is equally understood by biologists of different disciplines.

2.1 Various effects of weeds

Weeds may affect man's agricultural and non-agricultural activities in many different ways, some of which are outlined below:

(a) Weeds may be parasitic upon crop plants

This property is not of significance in Europe, though there are a few species of parasitic or semi-parasitic plants in the flora. Some parasitic weeds are extremely important in other parts of the world, however, examples being the species of *Striga*, the witchweeds, parasitic on sorghum and maize in Africa, India and parts of the United States.

(b) Weeds may be poisonous

Poisoning of humans by weeds is probably rather rare, but though grazing animals often avoid poisonous plants in pastures they do not invariably do so, and they may be unable to discriminate against them in hay or silage. Serious losses can occur. CRAFTS and ROBBINS (1962) quoted the annual losses of livestock in the U.S.A. from this cause at fifteen million dollars.

(c)   Weeds may be unpalatable, nutritionally poor or may cause tainting of animal products, even if they are not actually poisonous

The wild onion (*Allium vineale*), for example, causes unacceptable flavours in meat and milk. The palatability of hay and silage may be seriously affected by some weeds.

(d)   The physical characteristics of some weeds may be a problem

Plants with parts which become tangled in sheep's wool, such as the fruits of *Galium aparine* (cleavers) can be a serious nuisance and thorns or spines may cause injury to animals.

(e)   Serious infestations of weeds may cause damage to or at least interference with the functioning of farm machinery at harvest or other times

Knotgrass, which has long wiry stems which spread close to the ground, is a good example of a weed causing this sort of difficulty.

(f)   Even if the yield of a crop is not reduced its value may be seriously affected by the presence of weeds

This is especially likely to be true of crops grown for seed, where contamination by weeds greatly increases the cost of the cleaning operations needed. In some cases crops contaminated with more than a specified proportion of certain weeds are not acceptable as seed crops at all. Another example of this type of effect is the presence of black nightshade (*Solanum nigrum*) in crops of peas grown for the canning or freezing industries. The fruits of this plant are very much the same size and shape as peas, and as they are somewhat poisonous are an unacceptable contaminant of the harvested crop. This is a good illustration of a problem unique to a highly mechanized industry, since when peas were grown on a relatively small scale and sold in the pods there was no possibility of black nightshade's being a factor in the quality of the crop.

(g)   Weeds may act as hosts for diseases and pests which affect crop plants

A very small number of examples is given in Table 3. In this respect weeds are no different from other members of the wild flora of the area, but their close association with crops gives them particular importance.

(h)   Some specialized weeds may be important in blocking drainage ditches and irrigation channels (see TINKER, 1974)

In special cases, like that of the notorious water hyacinth (*Eichornia crassipes*), they may block major waterways.

(i)   Weeds may affect many man-made environments other than strictly agricultural land

Roads and railways provide good examples here, and quite large sums of money have to be spent to keep embankments and roadside verges reasonably clear. The plants of such environments are often the same as those of cultivated land, though there are some which cannot survive the

Table 3   Examples of weeds acting as hosts for diseases and pests of crop plants.

Type of pathogen or pest	Weed	Crop plant
Fungi:		
Claviceps purpurea (Ergot)	Black-grass	Rye
Ophiobolus graminis (Take-all)	Couch	Cereals
Plasmodiophora brassicae (Clubroot)	Various members of the Cruciferae	Brassicas
Viruses:		
Cucumber mosaic	Common chickweed	Many crop plants
Raspberry Ringspot (Nematode transmitted)	Common chickweed, creeping thistle and others	Raspberry, strawberry, redcurrant
Nematodes:		
Stem and bulb eelworm (*Dictylenchus dipsaci*) (Kühn)	Many weeds	Many crops
Insects:		
Black Bean Aphid (*Aphis fabae* Scop.)	Fat-hen, many legumes	Broad and field beans

very severe conditions of cultivated situations but which do very well in the slightly less stringent conditions of waysides and waste places. These are sometimes called *ruderals* and distinguished from the weeds of cultivation which are termed *agrestals*, but the distinction is not one which is called for if the definition of weeds by Baker, which was discussed earlier, is used. Biennials such as the white umbellifer *Anthriscus sylvestris* (cow parsley) which are not very common in cultivated land may be abundant on roadsides.

2.2   Interference with crop plants

In spite of the other important effects of weeds discussed above there is no doubt that the one of prime interest is the effect which they have on crop plants which leads to the reductions of yield, both in bulk and cash terms.

(a)   Interference with the crop in terms of space

It is a fact of biological life that if one organism occupies a particular space in an ecosystem then another organism cannot do so, and in the plant world space tends to be allocated on a strictly first come, first served basis. In this sense weeds might be said to compete with crops for a commodity which is in limited supply, but since the occurrence of a plant in a particular location may be due very largely to chance it might equally be argued that chance had pre-empted any actual competition for space. What is certainly true is that a weed occupying a space which might have contained a crop plant is extremely likely to be competing for other things too, and perhaps particularly so if it bears any general morphological similarity to the crop. Fig. 2–1 illustrates some aspects of this point.

(a)   **(b)**

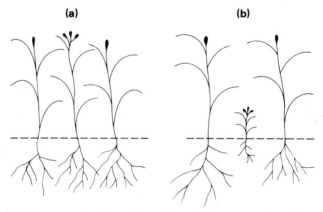

Fig. 2–1   Diagrammatic illustration of two different crop/weed situations: (**a**) weed (centre) resembling crop in general growth habit; (**b**) weed not resembling crop.

In case (a) the weed is of comparable stature to the crop and is occupying a place which might have contained a crop plant. Its roots and its leaves are distributed in a way similar to those of the crop and it seems likely that competition for light, nutrients and water may be taking place. In case (b), although the weed is in the same position in relation to the crop it is clearly not likely to be competing in the same way with the crop plants; these have, in fact, rather more space in which to grow than they would have if the space occupied by the weed were occupied instead by a fellow crop plant. This raises at an early stage the important concept, to which we shall be returning later, that crop plants do compete very considerably with each other. Another point relevant to Fig. 2–1 is that, although interference with crops by weeds is likely to be particularly severe if there is a morphological similarity between the two this similarity is by no means a condition for successful interference by a weed. The weed in (b) may in the earlier stages of growth have had an effect on the crop which will last until harvest, and provided it is able to flower and set seed when overtopped by the crop its successful weediness is not in dispute.

There are many cases in which highly successful weeds do show morphological and physiological resemblances to crops. Examples often quoted are those of wild oats in cereal crops, *Camelina sativa* (gold-of-pleasure) in flax, and grass weeds in pastures. The important points of resemblance may in other cases be restricted to things such as the size and shape of seeds, but it is important to bear in mind that associations between particular weeds and particular crops are likely to depend as much, if not more, on the ability of the weed to survive the system of cultivation used for the crop as on any other factor. A good example of this is given by two weedy species of *Avena*. The wild oat (*Avena fatua*) germinates largely in the spring, whilst the winter wild oat (*Avena ludoviciana*) germinates in the autumn and winter. The latter is able to become established in crops of wheat sown in the autumn since it can produce seedlings during the period before the wheat has reached a size at which it is an effective competitor. On the other hand the cultivation of land for the sowing of spring wheat destroys seedlings of the winter wild oat, and even on land heavily infested with seeds of this weed it is possible to grow crops of spring wheat completely free of it. The reverse story is true of the spring wild oat which is a much worse competitor with spring than with winter wheat, because the latter is already well established by the time *Avena fatua* seeds are ready to germinate.

(b)   Interference in terms of water, nutrients and light

These factors are considered together, at least as a heading, for the very good reason that it is often difficult or impossible in practice to separate their effects. For example, in any situation where nitrogen is in short supply leaf expansion in both crops and weeds will be reduced and this will then affect the extent of mutual interference in terms of light. Similarly plants which root at different depths may have access to different amounts of water, but nutrient availability will also be different. BLEASDALE (1960) in making the same point suggested that the overall growth of the plants involved in a competitive situation, measured in terms of weight, could be regarded as integrating the various components of the competitive process. In complex situations this is a more useful way of considering the position than trying to separate the different components from each other.

The question of competition for water is one which presents special problems, partly because the rate of loss of water from land is governed much more by the environmental conditions than by the type of plant growing on the land, provided that certain other conditions are fulfilled. None the less weeds clearly may have an influence on crops through their uptake of water; the nature of the influence will depend on the relative rooting depths of the weed and the crop. In addition in areas of low rainfall a substantial cover of weeds may prevent a large proportion of the rain falling in a fairly light shower from reaching the ground at all. The situation in relation to nutrients is rather less difficult to understand, since at any one time there is a finite and perhaps an inadequate quantity of a particular element in the soil and competition, in the strict sense, can certainly take place for this limited supply. In so far as ability to reach and obtain nutrients which are present in the soil is likely to be an important factor in the success of a plant establishing itself in an open situation, there are two characteristics which successful weeds might be expected to possess. One of these is the ability to produce an ample root system and to do so quickly, and the other is to be efficient in the accumulation of necessary nutrients. In some painstaking experiments in 1934 PAVLYCHENKO and HARRINGTON measured the total root length of a number of weed and crop species 5 and 21 days after germination. They showed that there were very marked changes in the amount of root produced amongst both classes of plant but that many of the crop plants had more roots than the weeds five days after germination. In some ways this is not surprising, since one of the characteristics for which man has both consciously and unconsciously selected his annual crop plants is rapid and uniform germination and establishment. After 21 days, however, many of the serious weeds amongst those tested, including fat-hen and charlock, had root systems at least as extensive as those of the crops. The significance of root systems which may explore different layers of the soil has already been touched on.

In relation to the efficiency of nutrient uptake there is quite a lot of information available which suggests that some weeds are particularly effective in the uptake of particular nutrients. BUNTING (1960) points out that where soil has been disturbed, there is an initial flush of nitrogen due to the effect of the supply of air made available by the disturbance on the soil nitrifying bacteria, but that in the secondary succession which follows, nutrients may be quickly leached away before a closed vegetation cover is achieved. Having evolved under such circumstances weeds may consequently be expected to show a high ability to take up available nutrients. As an example he quotes the tropical weed *Tridax procumbens* which appears to be a very efficient accumulator of phosphorus, an element often lacking in tropical soils and one which, even when present, is frequently immobilized by adsorption in the uppermost layers of the soil. In a similar way charlock appears to be a good accumulator of nitrogen and fat-hen shows a marked ability to accumulate potassium. In the experiments of BANDEEN and BUCHOLTZ (1967) on nutrient competition between *Agropyon repens* and maize it was shown that substantial amounts of nitrogen, phosphorus and potassium were taken up by the common couch, especially early in the season. For example, in July the couch shoots and rhizomes contained 68% of their total uptake of potassium for the season. Since maize only starts its major period of growth in late June it may well have suffered from the reduced potassium concentrations caused by the weed. Despite the evidence which is available on this subject there does appear to be a shortage of really critical data on competition for nutrients as a factor in interference with crop growth by weeds and especially on the question of whether the ability of some weeds to take up specific ions particularly effectively is an important factor in their competitive ability. One interesting observation relating to this is that black-grass (*Alopecurus myosuroides*), a serious annual grass weed of cereals on heavy land, appears to have an unusually low requirement for potassium, and this may explain its ability to compete with wheat on soils low in potassium.

That ability to compete effectively for light is an important factor in plant competition is virtually a dogma in plant ecology. There is no doubt that some weeds show a mode of growth which makes them very effective in this way. Conspicuous examples of this are the dicotyledonous weeds of lawns or grazed pastures such as plantains, daisies, dandelions and docks, which share a mode of growth in which wide leaves can spread out relatively close to the ground. Heavy infestations of such annual weeds as fat-hen or common chickweed, one of which can grow fairly tall and the other can scramble over other plants, can reduce the light levels close to the weeds very considerably, to the undoubted detriment of the crop, but in many experiments the effects of competition for light are undoubtedly complicated by competition for other environmental factors. Illustrations of some of the studies which have been made on weed/crop competition are given in the next section, but in these no attempt is made to distinguish the various components of the complex situations involved.

(c)   Some experiments on weed/crop interference

Much of what has been said earlier has suggested that it might be during the very early stages of crop growth that weeds may be particularly effective, and many experiments bear this out. SHADBOLT and HOLM (1956) studied the effects of weeds on several vegetable crops and showed that in many cases the first four weeks were critical. Interference with growth by weeds at this time often produced effects which were still visible at the time of harvest. The extent of this effect depends on the crop. DAWSON (1964) working with *Phaseolus* beans showed that keeping the crop weed-free for 2, 3, 4 and 5 weeks after sowing caused a steady increase in yield, but there was little effect of weeds which established after this time (see Fig. 2–2). Several authors have suggested that in a relatively short season crop it is the first 30 days or so which represent the really crucial period for the development of effective competition.

(a) **(b)**

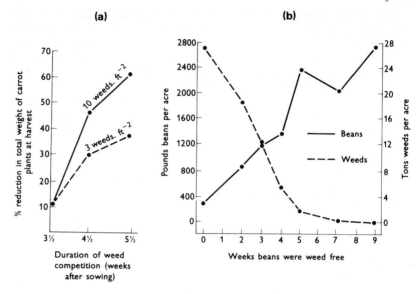

Fig. 2–2 Results of two experiments on the interference caused by weeds with the growth and yield of a crop. (**a**) Effect of duration and extent of weed competition on weight of carrot plants at harvest. (Data of SHADBOLT and HOLM, 1956.) (**b**) Effect of length of weed-free period on yield of threshed beans and on weight of weeds at harvest. (Redrawn from DAWSON, 1964.)

A distinct complicating factor is that in experiments in which a given crop is grown in competition with different weeds the response depends upon the species of weed. For example BLEASDALE (1960) reported on experiments in which carrots were grown in competition with three different annual weeds at two different levels of soil fertility. *Poa annua* did not reduce carrot yield at either level of nutrition, but chickweed and groundsel did. In the case of chickweed the yield of carrots in the pots with the weed at high levels of nutrients was the same as that in the pots without the weed at low levels, suggesting that in this case competition for nutrients might be an important factor. In an interesting experiment with beetroot Bleasdale showed that where weeds were left to grow *in* the rows of the crop but removed in the area *between* the rows the yield of beet was greater than when the reverse treatment was applied. The crop competed more effectively with weeds which were very close to it than those a little further away. This also suggests that perhaps soil nutrients might have been depleted by the between-row weeds more than by the within-row ones, which would themselves have been subject to much more serious interference by the crop. Beets of all kinds, once well grown, are good competitors for light because of their rosette of large leaves.

As a final example of the complexities of a situation when more than two plants are involved it is interesting to look at the work of HAIZEL and HARPER (1973). They grew plants of barley (*Hordeum sativum*), white mustard (*Sinapis alba*) and wild oat (*Avena fatua*) together in all combinations at different densities and included treatments in which competing species were removed after various time intervals. The design of the experiment enabled them, amongst other things, to rank the three species in order of effectiveness as competitors against each other and against themselves, and the results obtained are summarized in Table 4.

Table 4　Three species ranged in order of aggressiveness from experiments in which all combinations of pairs were grown, including the effect of intra-specific competition in each. (From the data of HAIZEL and HARPER, 1973.)

In competition against	Most aggressive	Next most aggressive	Least aggressive
Barley	Barley	Wild oats	White mustard
Wild oats	Barley	White mustard	Wild oats
White mustard	White mustard	Barley	Wild oats

Two very important features emerge from this part of their results. One is that the apparent relative aggressiveness (or competitive ability) of a plant in relation to others depends entirely on what plant it is competing with. This means that it is very unlikely that one could ever list weeds in a 'league table' of competitive ability. The second feature of interest is that both barley and white mustard show their greatest competitive ability against themselves—that is intra-specific competition is stronger than inter-specific competition. This is a result which would be expected by ecologists and evolutionists on the basis of Darwinian theory, but wild oat provides an exception in being a stronger competitor against the other two species than it is against itself. The same is true of the annual grass weed black-grass (*Alopecurus myosuroides*) in some experiments. These two grass weeds also share the ability to cause decreases in yield of cereals even when present at very low densities, though discussions of density effects in weeds are beset by the difficulty that similar yield reductions can be obtained in some cases from a wide range of densities; in these instances such factors as relative time of emergence of the weed and the crop may be very important.

(d)　Interference with crops by means of toxic exudates

There is no doubt that many plants do produce chemicals which may be released from roots or leaves and which will, under certain conditions, adversely affect the growth of other plants. The exact ecological significance of this phenomenon, however, is one which causes much controversy. Examples of the sorts of experiment which suggest that such a direct form of influence of one plant on another may take place are given by MARTIN and RADEMACHER (1960), GRÜMMER and BAYER (1960) and WELBANK (1960, 1963). Martin and Rademacher's experiments were carried out with plants grown in nutrient solutions with the pots banked so that solutions could be circulated without roots ever coming into contact with each other (Fig. 2–3).

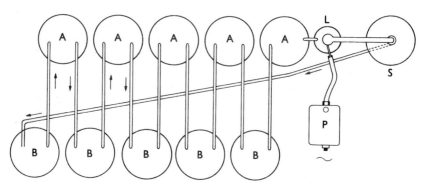

Fig. 2–3   Diagram of the layout of the experiments of MARTIN and RADEMACHER (1960). Two series of 4 litre culture vessels were connected together as shown so that the culture solution flowed through alternate vessels containing either species A or species B. The solution was kept in circulation by means of the pump P and the lifting apparatus L. The vessel S was a reservoir not containing plants.

They found that several plants grown in this way inhibited each other's growth, but the presence or absence of the effect depended upon the species used—some had no effects at all. It is, however, always extremely difficult to extrapolate from the effects of experiments in nutrient culture to the situation in the field. Grümmer and Bayer worked with flax and a well-known weed of this crop *Camelina alyssum* which can cause serious losses of yield even when present in relatively low densities. Their results, which do not appear to have been confirmed by other workers up to now, showed that there was no effect of the weed on the crop when both were grown together in pots, provided the pots were watered from below so that the water did not fall on the leaves. This was true whether or not the roots of the weed and the crop were allowed to mix in the pot. When the pots were watered by a spray from above, however, serious interference with the growth of the crop was caused. This appeared to be due to the leaching of toxic materials from the leaves of the weed by the water, and two of the toxic constituents of these leachings were chemically identified as phenolic compounds, many of which are notable for their growth inhibiting properties.

Welbank worked with *Agropyron repens* which has often been considered to have a toxic effect on other plants. He found from various experiments that nothing produced by the roots or rhizomes inhibited the growth of several test species, even when these were grown in close contact with the weed roots. When extracts made from the ground up residues of roots or rhizomes were placed in soil they inhibited the germination and growth of rape seedlings very strongly, suggesting that chemicals produced by the breakdown of the underground parts in the soil might be involved in the competitive effect. These experiments have been criticized on the grounds that the effects might be due directly or indirectly to microbial action in the soil as a result of the residues incubated with it, but more recent evidence has again suggested that the effect observed may be a real one though its mechanism is not yet known. Grass weeds other than *Agropyron* have been shown to exhibit similar effects, and the whole subject still requires further critical study. There is nothing intrinsically unlikely in the concept of the metabolites of one plant having an adverse effect on the growth of others, and the possession of such a mechanism by a weed would be one more factor contributing to its success, even if the response to the toxin were confined to a limited number of its competitors.

Fig. 2–4 gives a summary of some of the factors which are involved in the mutual effects of crop and weeds.

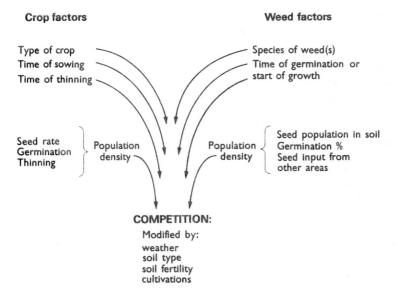

Fig. 2–4   Some of the factors influencing the degree of mutual interference, or competition, between weeds and crops.

3 Characteristics of Weeds. 1

Many of the characteristics which are likely to make a plant into a 'good' weed are almost self-evident. At the very least it is possible, without straining credulity to any real extent, to draw up a list of characteristics which are likely to be valuable assets to such a plant even before looking carefully at weeds to see if they really do possess such characteristics. An instructive exercise might be to do this now, before reading any further, to see how many features of a list thought up on this basis recur in the discussion in the next two chapters. This exercise might well be called 'design your own weed'.

No one weed is likely to possess all of the characters in any such list, and one can only be grateful for this! In some cases characters which seem very important may be conspicuous by their absence. None the less, when comparing any named weed against a list of possibly desirable weedy characteristics we normally find that many of these do occur. Some characteristics may be associated with each other—for example very large seed numbers are frequently associated with a small seed size—but others may be thought of as having simply come together in one plant under the influence of the selective pressures which have led to the evolution of weeds.

The list of characters which will be discussed in a certain amount of detail in this chapter and in chapter 4 is based on, but not identical with, that given by BAKER (1965). In this chapter we shall look at some of the characteristics which may be particularly applicable to annuals (though not exclusively so) since they are associated especially with seeds and seedlings. In chapter 4 we shall look at some more general features and some which are specific to perennial weeds. In chapter 4 some biological adaptations to seed and fruit dispersal are discussed, but other aspects of this matter and a more detailed consideration of some genetical and evolutionary aspects of weeds are postponed to chapter 5.

The following possible characteristics of an 'ideal' weed are therefore dealt with in this chapter:

High output of seeds in favourable conditions.
At least some output of seeds even in very poor conditions.
Seed production begins after only a short period of vegetative growth.
Seed production spread over a long period of the growth of the plant.
Variable seed dormancy and considerable longevity of seeds in soil.
Has no special environmental requirements for germination.
Rapid seedling growth and establishment.

3.1   High output of seeds in favourable conditions

Other things being equal, a species able to produce a large population quickly will be at an advantage in a competitive situation, and many weeds are notable for producing large numbers of seeds. Frequently they are able to do this in the absence of cross pollination (see chapters 4 and 5) which is clearly an advantage in species colonizing new areas. Table 5 gives data which emphasize that seed production is very high even in the less prolific of the common weeds listed.

Table 5   Average output of seeds or fruits for ten common weeds (data of SALISBURY, 1961). The asterisk indicates perennials.

Weed	Average output per plant	Seed or fruit
Groundsel (*Senecio vulgaris*)	1000–1200	F
Common chickweed (*Stellaria media*)	2200–2700	S
Shepherd's purse (*Capsella bursa-pastoris*)	3500–4000	S
Greater plantain (*Plantago major*)*	13 000–15 000	S
Common poppy (*Papaver rhoeas*)	14 000–19 500	S
Prickly sowthistle (*Sonchus asper*)	21 500–25 000	F
Perforate St John's wort (*Hypericum perforatum*)*	26 000–34 000	S
Canadian fleabane (*Conyza canadensis*)	38 000–60 000	F
Hard rush (*Juncus inflexus*)*	200 000–234 000	S

Comparisons of seed or fruit number per plant are clearly only of limited value, since individual plants vary greatly in size, and in any case it is not only the number of seeds produced but the number germinating and producing seedlings which is really critical. The production of very large seed numbers tends to be accompanied by small, light seeds which are at an advantage in short distance dispersal under the influence of wind even if they are not specially modified for wind dispersal, as some are (for example, many members of the Compositae). It has been pointed out, however, that very small seeds do have serious disadvantages when it comes to germination and establishment, and in drier areas the possession of large seeds may be a distinct advantage in this respect. That very high seed production and small seeds are not a condition of success in an annual weed is shown by the sunflower (*Helianthus annuus*), a very successful wayside plant in the United States, which has conspicuously large and heavy seeds. Perennial plants are not under the same pressures towards high seed production, and some are certainly less prolific than many annuals, but this is far from universal as many of them also produce large numbers of seeds (see Table 5).

3.2   Seed output in poor conditions

Plants in the types of habitat frequented by weeds are likely to encounter very poor conditions for growth fairly frequently, whether this is due to the exposed conditions, to human interference or to competition from crop plants. In such conditions it is crucial from the point of view of the survival of the species that an individual should produce at least some seeds. It is not surprising, therefore, to find that many annual weeds can produce at least a few seeds, even on very small plants. HARPER (1960) quotes a range of seed production observed in the common poppy from one capsule containing only 4 seeds up to 400 capsules each containing 2000 seeds. Similarly THURSTON (1972) indicates that a plant of black-grass may have a single inflorescence and produce 5–10 fruits or may have up to 200 inflorescences with 100 fruits in each. A related phenomenon from the point of view of species survival under cultivated conditions is that in many cases seeds of weedy plants may be viable before full maturity, so that if the weed is cut down before the normal maturation period is complete at least a proportion of seed production is not lost. Examples of this feature are provided by wild oat, groundsel and ragwort.

The important point about the flexibility of seed production shown by many weeds is that it is a feature which not only tends to ensure the survival of the species (one of the 4 seeds in Harper's poppy plant mentioned above might give rise to a plant with the characteristics of his second plant in another year) but that it tends to make weed populations very resistant to change in situations where a species is competing with itself. In experiments with corncockle sown at different densities HARPER (1960) showed that a reduction of the seedling population by 90% only reduced the seed production of the population by 10%. This observation has obvious lessons for those concerned with weed control by cultivations, especially under fallow conditions, since it shows that it is not enough simply to reduce a population, even if the reduction is drastic, unless the plants remaining can be prevented from seeding.

3.3   Seed production and the age of the plant

A corollary of 3.2 above is that under adverse conditions there is much to be said for being able to flower and produce seeds at a very early stage of growth, and a very few observations of plants like shepherd's purse or annual meadow-grass growing on footpaths will show that this does indeed happen. (See Fig. 3–1.) This is not a characteristic of all weeds, but many of the common annual weeds show it, and it is a character which contrasts strongly with most crop plants, and is closely related in this discussion to 3.4 below.

Fig. 3–1 Two common annual weeds, illustrating the ability to flower at different stages of growth. (**a**) Groundsel, (**b**) Shepherd's purse.

3.4 Spread of seed production over the growth period of the plant

Even if seed production can begin very early in the life history of the plant, it might be considered an advantage to be able to continue producing flowers and seeds for a fairly long time, if conditions permit. This is a character very well shown by some weeds and very much less so by others. Charlock tends to produce most of its flowers and fruits in one burst, whereas the small nettle (*Urtica urens*) produces seeds over a very large proportion of its growing period. This character is not to be confused with that of all the year round seed production by weeds like annual meadow grass; this is simply due to the fact that seed germination and seedling growth go on throughout the year so that it is always possible to find plants in flower irrespective of the actual length of the flowering period of any individual specimen.

3.5 Seed dormancy and longevity

The ability of weed seeds to germinate after variable and often long periods of dormancy is one so vital to their success that there is an enormous amount of data available on it. In showing the phenomenon of seed dormancy so commonly weeds may be contrasted with most crop plants, which have been selected both consciously and unconsciously by man for rapid, uniform and immediate germination without any dormant period. Such behaviour in a weed would mean that it could be eliminated entirely by only one or two years of careful cultivation.

Dormancy in seeds is often considered as falling into three categories; innate, induced and enforced. Innate dormancy is genetically determined and means that a weed will not germinate under any conditions for a period after it has been shed from the plant. An enormous number of weeds have seeds falling into this category. Induced dormancy is a condition in which dormancy is caused by the particular conditions to which the weed is exposed and in which it continues even after the conditions which gave rise to it are changed. High concentrations of CO_2, for example, will induce dormancy in some seeds. Since high CO_2 concentrations are often found in the soil, seeds with this characteristic would be likely to become dormant if buried and not to germinate unless brought near the surface where the excess CO_2 could diffuse away. The advantage of this to a weedy species whose seeds are likely to be buried by cultivations (especially by ploughing) is self-evident. In enforced dormancy seeds do not germinate simply because the environmental conditions are not suitable. A seed with a light requirement for germination (a characteristic of many weeds) will show enforced dormancy if buried even at a relatively shallow depth.

Many weed species show a marked periodicity of germination. Under natural conditions most species have a period or periods in the year when they germinate best (see Fig. 3–2), and in the experiments of Brenchley and Warington in the 1930s in which soil was kept in a greenhouse, stirred regularly to bring new seeds to the surface and germination recorded regularly, this behaviour continued over a period of three years. Similar results have been obtained by other workers with species different from those used by Brenchley and Warington.

It is an interesting sidelight on the question of seed dormancy in weeds that in experimental work, for example on the screening of chemicals for herbicidal activity, special steps have to be taken to break the dormancy of many seeds so as to give adequate populations of plants for testing. Excellent accounts of many aspects of the subject of dormancy in weed seeds are available (e.g. THURSTON, 1960; SALISBURY, 1961) and these contain many references to experiments which have been carried out in this area.

The question of the longevity of seeds in the soil is clearly a closely related one. Most crop seeds, and those of many wild plants, have a relatively short life when buried. In the case of weeds, however, considerable longevity is a widespread characteristic. This might be anticipated for any plant whose habitat is such that its seeds are apt to be buried up to 18 inches deep and perhaps not brought to the surface again

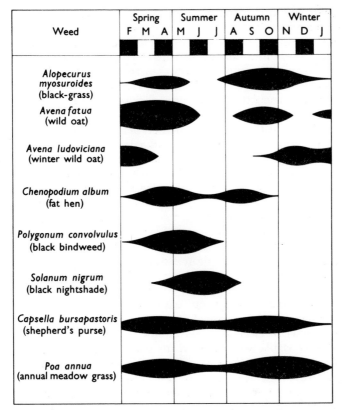

Fig. 3–2 Diagram showing the main germination periods of eight important weeds. Several different seasonal patterns are represented. (Modified from ROBERTS, CHANCELLOR and THURSTON, 1968, with permission.)

for several years. The two classic studies on the subject which are discussed in many accounts are those of Beale and Duvel. Both men buried samples of seeds and recovered them at long intervals for germination tests. (It would be better to say 'arranged for them to be recovered', since Beale's experiments have now been running for more than 90 years!) The precise details of these experiments are not significant here, but the general conclusions are clear. Many weed species showed large percentages of germination even after long periods, e.g.

Shepherd's purse	47% after 16 years
Greater plantain	84% after 21 years
Black nightshade	83% after 39 years

(Duvel's experiment. Reported by TOOLE and BROWN, 1946.)

Information more closely related to the field situation can be obtained from studies of the weed seed populations of areas which have been under permanent pasture for known periods. Information provided by samples from such areas confirms that very many weed seeds remain alive for very long periods and that cultivation of a 5- or 10-year-old pasture which was formerly arable land will turn up enough viable seeds of arable weeds to re-infest arable land if the seedlings produced become established. This is not to suggest that weed seed populations of soils do not decline with time, since they clearly do. The immensely detailed and thorough studies of Roberts over many years have shown that total viable seed populations may fall fairly rapidly if land is maintained weed-free (e.g. ROBERTS, 1958, 1970). The fact is, however, that because of the large reproductive potential of many weeds it does not need more than a relatively small population of a weed one year to lead to a serious infestation the next. Two interesting conclusions derived from the Duvel experiment which are relevant here are, firstly, that seeds buried more deeply survived better than those at shallower depths, and secondly, that weeds characteristic of an area survived better than those from other areas. The advantage of better survival at greater depths is clear when relatively deep ploughing is practised, and may well be due to the more stable environment at greater depths. Temperature fluctuations and changes in the gaseous environment are both less at greater than at shallower depths.

From the ecological and agronomic points of view the actual population of viable weed seeds in an area is obviously of interest. Fig. 3–₃ summarizes some of the factors which may influence this. Detailed

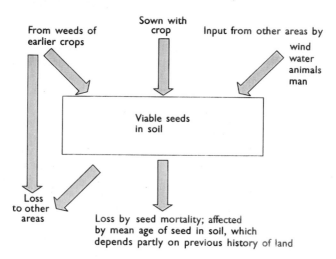

Fig. 3–3   Diagram illustrating some of the factors influencing the total population of viable seeds in the soil.

information is available about some of these factors, at least for some weeds, but little is known of others, and some may vary widely from area to area (see chapter 5 and HARPER, 1960). The number of seeds found in the soil is, however, often extremely large. Table 6 gives a few examples.

Table 6   Populations of viable weed seeds of some arable soils in Britain. (From ROBERTS, 1970, slightly modified. The original references may be found in this paper.)

Source	History of land	No. of seeds, m^{-2}
Rothamsted	Continuous wheat	34 100
Woburn	Continuous wheat	28 700
Woburn	Continuous barley	29 900
Wellesbourne	Mainly cereals	56 600
Midland clay	Farm crops	4000–27 400
Derelict arable	Farm crops	4900–46 200
Old arable	Farm crops	250
Experimental	Vegetable crops	3800–15 300
Commercial	Vegetable crops	1600–86 000

In any one year it is the weeds nearest to the surface which are likely to be important, since those lower down will not germinate or, if they do, will stand less chance of reaching the surface. HITCHINGS (1960) showed in one experiment that in the case of common chickweed, of a population of 291 seeds per 9 inch square, 254 were in the top 3 inches. This gave rise to an average population of 12 plants per square yard. This is not a dense population, but could easily lead to one, as is recognized by the old gardeners' saying 'one year's seeding, seven years' weeding' which takes account both of the prolific production and of the longevity of seed of weedy species.

3.6   No special environmental germination requirements

This possibly desirable attribute for a weed species from Baker's list is included in this form because when considered initially it seems reasonable. Plants with special germination requirements are likely to be more restricted in occurrence than those not needing particular combinations of conditions. Further thought, however, soon leads to the opposite conclusion, as has been mentioned earlier (section 3.5). Weed seed germination has to give populations competing successfully with crop plants, and this may be done more effectively if it is closely related to some factor which links it to cultivation operations. In chapter 2 the example of winter wild oats was quoted; here autumn and winter germination make this plant an insignificant weed in spring cereals.

Spring wild oats show a period of dormancy and a period of cold is a factor in terminating this, leading to germination in the spring and consequently to successful competition with spring sown cereals.

In the case of crops uniformity of germination and seedling establishment are important, and much attention is therefore given to the provision of a uniform seed bed. In weeds this is much less critical, since provided there is an adequate supply of suitable sites for germination the distribution of these is not important. In fact the actual requirements for germination of a particular species may be relatively precise, and Harper uses the term 'microsite' to describe the particular location offering the right germination conditions for a given species in a relatively heterogenous environment. However rugged or unsuitable most of the environment may be, it is the frequency of appropriate microsites which is important from the point of view of any one species.

The periodicity of germination of many weeds has already been mentioned and illustrated in Fig. 3–2, and in itself indicates that there may be special requirements for germination. Even when conditions are appropriate, however, not all species behave in the same way, as some show very rapid and almost synchronous germination of all the seeds whilst others show germination over a period. Fig. 3–4 illustrates this. A third type of behaviour, that of intermittent germination over a rather longer period, is presumably due to the variable dormancy of some species, since dormancy is by no means always lengthy. Even species germinating at all times of year frequently show this intermittent pattern of germination, which seems very likely to confer an advantage on a plant colonizing an open and rapidly changing habitat. A phenomenon found

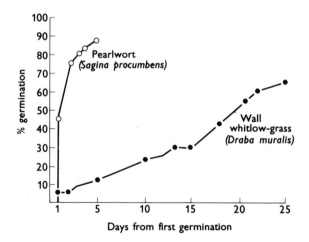

Fig. 3–4 Graphs showing two different patterns of germination as demonstrated by pearlwort and wall whitlow-grass. (From SALISBURY, 1961.)

in some plants which is associated with this variable pattern of germination is an actual polymorphism of the seed or fruit. Different visible features of the propagule may be associated with slightly different germination behaviour.

The need for light to stimulate germination in many species of weed is notable, but this is a physiological requirement which may disappear as the seeds age, though it does not necessarily do so. In the experiment of WESSON and WAREING (1967) the weed seeds studied were in soil from a field which had been under continuous pasture for at least six years, and very many species germinated only in response to light.

A very interesting fact about the crop/weed relationship is that the presence of a crop may actually improve the conditions for weed establishment after germination, presumably by reducing the extremes of temperature and the drying effect of wind experienced in the absence of the crop (see HARPER (1960), p. 128). The balance of advantage between this and the importance of an early start in the competitive situation between crops and weeds presumably varies according to the particular crop/weed species involved and the other conditions.

Another example of a special germination requirement which is relatable to an existence in an open environment is that of the common poppy, whose seeds germinate better if exposed to a wide fluctuation of temperature between day and night.

A spectacular example of the possible importance of the precise environment of the seed for germination is given by the experiments of HARPER et al. (1965) (also cited by HARPER, 1965) on two *Plantago* species. These are illustrated in Fig. 3–5 which shows the distribution of seedlings of *P. media* and *P. lanceolata* on an area of soil in or on which various modifiers of the environment were placed, or where the soil surface was depressed to form square hollows. *P. media* showed particularly marked germination beneath sheets of glass (treatment 3); *P. lanceolata* showed germination in a wide variety of situations but this was particularly marked around the edges of obstructions or depressions, whose shapes can be clearly seen from the pattern of seedling emergence.

Two further pieces of work may be quoted to indicate some evolutionary aspects of special germination requirements. In one of these CUMMING (1963) showed that non-weedy species of *Chenopodium* did in fact have a narrower range of conditions for germination than is known to be the case for the weedy species, fat-hen. Thus, although weed species may have special germination requirements they must not be thought of as necessarily more exacting than those of non-weedy relatives.

In the second study (HARPER and McNAUGHTON, 1960) it was shown that hybrids between *Papaver* species had seeds which lacked the dormancy shown in the parental species (about one sixth of the seeds in a capsule of *P. rhoeas* show some dormancy, the rest being capable of immediate germination). This loss of a physiological requirement in the hybrid

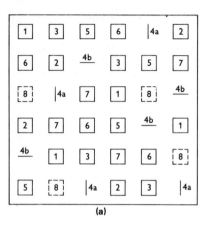

(a)

Fig. 3–5 The distribution of seedlings of *Plantago media* and *Plantago lanceolata* in relation to the placement of various objects on or disturbance of the soil surface. (**a**) Arrangement of objects and disturbances (all the small squares represent areas of 144 cm²), (1) Soil depressed 1.3 cm. (2) Soil depressed 2.5 cm. (3) Glass plate laid flat on soil surface. (4a) Glass plate placed vertically on soil, 2.5 cm below and 1.3 cm above surface, aligned north–south. (4b) As 4a, but aligned east–west. (5) Open wooden enclosure projects 2.5 cm above soil surface. (6) As 5, but projects 1.3 cm. (7) As 5, but not projecting. (8) No treatment. (**b**) Distribution of seedlings of *Plantago media*. (**c**) Distribution of seedlings of *P. lanceolata*. (From HARPER *et al.*, 1965; redrawn with minor omissions.)

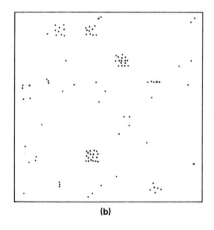

(b)

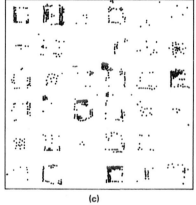

(c)

would have important consequences in evolutionary terms, since it would largely eliminate the hybrid as a competitor with either parent unless the seedlings were particularly hardy. It would be interesting to see if similar cases could be found in other genera where formation of fertile hybrids is possible; the physiological background of this type of behaviour is of great interest, irrespective of its significance in the evolution of particular species.

3.7   Rapid seedling growth and establishment

The importance of rapid and effective establishment of seedlings has already been mentioned a number of times, particularly in chapter 2. This is probably the crucial stage in the life of any plant, whether weed or not, and crops have certainly been unconsciously selected for ability to establish well, since those which do not will be rapidly swamped by weeds. The young seedling stage of weeds is one to which much attention has been given by herbicide workers, since it is at this period that most weeds are particularly sensitive to herbicides.

It is not vital that every weed plant should be a large and vigorous one before the crop grows enough to compete effectively, since in some cases conditions for the weed may improve again late in the season when the crop leaves die off or the crop is harvested. Provided the weed seedlings had established well enough to survive the period of strong crop competition they might then be able to produce new growth late in the season. Interesting light is thrown upon the importance of seedling establishment by the experiments of SAGAR and HARPER (1960) with plantains. Seeds of three *Plantago* species were sown in a very wide range of environments, in some of which no plantains were found naturally. Young plants were also transplanted into the same conditions. In the case of the transplants all remained alive for at least a year except in two environments where light was very limited due to the presence of other plants. In a number of the same habitats the plants grown from seed did not survive; this emphasizes the sensitivity to the stress of external factors of the period following germination. This being so, any characteristic of a species which lessens its sensitivity to competition or to physical factors in the environment is likely to have a very strong survival value and give its possessor an important competitive advantage.

4 Characteristics of Weeds. 2

In this chapter the weed characteristics considered are those likely to be an advantage to most weeds, but not specifically associated with the seeds or seedlings of annuals. They are as follows:

Wide tolerance of variations in the physical environment.
Adaptations for both long and short distance dispersal.
Good powers of vegetative reproduction and ability to regenerate when divided into fragments.
Self compatibility.
Strong competitive ability.

4.1 Tolerance of variation in the physical environment

The physical environment of weeds may at various times be distinctly hostile. The soil surface is frequently both fully exposed and modified by cultivation, and this leads to conditions where (a) extremes of temperature may be experienced, (b) evaporation rates may be high, leading to serious shortage of moisture in the surface layers, and (c) the surface may well be puddled by heavy rain. If this happens it leads to a lack of penetration by water and to considerable run-off. Plants able to tolerate such conditions, especially as young seedlings, are likely to be good weeds. The characteristics mentioned are also found to varying degrees in other open habitats, except that in many of these the physical disturbance of the soil surface is absent. This disturbance has a number of effects which are not immediately obvious. One of these is that it leads to an annual flush of nitrates in the soil due to the high activity of nitrifying bacteria stimulated by the high concentrations of oxygen thus provided. Tolerance of this temporary period of high nutrient supply, followed in some cases by a more or less extended period of nutrient deprivation, is necessary for weeds. In areas where fertilizers are extensively used nutrient deficiency may be relatively rare, but in many parts of the world it may well be the normal state of affairs, and the ability to grow and reproduce in such conditions is vital to the successful weed.

The tolerance of variations at the local level must be extended to embrace other factors of the environment if a particular weed is successfully to make the transition to a totally new area—even a new continent. The cosmopolitan nature of many weeds has been mentioned before, and it is clear that very few indeed would have been successful on a worldwide or even a more limited scale if, for example, their flowering requirements were rigid. The work of CUMMING (1963) showed that several non-weedy species of *Chenopodium* had more specific requirements for flowering than *C. album*, and a similar trend in relationships between weedy and non-weedy species in the same genus is apparent in the genera *Ageratum* and *Eupatorium* (see chapter 5, Table 11).

Many plants, if not most, have adaptations in their life cycles which enable them to avoid the consequences of a season unfavourable to active growth. Weeds may have to survive two such periods, one climatic (low temperature or drought) and the other due to the acute stages of competition with crop plants. Seeds are of course an ideal vehicle for overcoming the climatically unfavourable period; the other may be particularly effectively avoided by perennial weeds. These may be able to make strong growth in the period after the maturity and harvest of a crop, though this ability is by no means confined to perennials.

It is difficult to generalize about the question of tolerance to environment, but there seems no doubt that it must be regarded as one of a weed's most important assets, and the geneticist Sakai has suggested that selection for this ability may well have been more important in weed evolution than selection for ability to compete with other plants (SAKAI, 1961).

4.2   Adaptations for long and short distance dispersal

The ability to produce offspring which may be established at some distance from the parent plant is a characteristic of fundamental importance to all plants, and perhaps especially to those of harsh or rapidly changing environments and with the ability to colonize new areas. Many weeds share the dispersal mechanisms common in the families to which they belong, but it is obvious that effective dispersal by natural means is not an absolute requirement for success as a weed. As will be discussed in more detail in chapter 5, man and his activities are the major disposal agents for weeds, but natural agencies and appropriate modifications of the plant which aid dispersal may be important also. The parachute-like pappus of the Compositae, for example, may enable seeds of weedy members to be spread over relatively long distances. Both docks and sedges have fruits which float easily and are readily dispersed by water, and adaptations causing seeds to be distributed attached to animals are also common (e.g. *Galium aparine*). The ability to germinate after passage through the alimentary canal of animals or birds may also be significant, and this has been found in black-grass, shepherd's purse, common fumitory, plantains, annual meadow-grass, and common chickweed, amongst many others. Cases of germination actually improving after such passage through an animal have been recorded, though these are probably rare.

As a cautionary note that, in the case of dispersal, things are not always what they seem, the work of BAKKER (1960) is interesting. Bakker studied the colonization of recently drained polders in Holland, by colt's-foot

(*Tussilago farfara*) and creeping thistle (*Cirsium arvense*). Both are members of the Compositae and both appear on casual observation to be most efficiently distributed by wind. This indeed was the case with colt's-foot, which was carried up to 4 km by wind and whose fruits germinated on the surface of the reclaimed polder. Only an extremely small number of the wind blown pappuses of creeping thistle turned out to have the achene attached (0·2% at a distance of 1 km from the source) and in any case germination of this species was better when the fruit was buried than when it simply lay on the surface. The study suggested that initial colonization of the area by colt's-foot was a result of its effective dispersal and appropriate germination requirements but that the subsequent arrival of creeping thistle was due to introduction with crop seeds. The question which needs to be asked, therefore, in considering natural adaptations for dispersal in weeds is not so much whether adaptations are actually present, but to what extent they are important to the weed and its mode of life. It may well be that one of the most important adaptations is in fact the seed dormancy and longevity discussed in the previous chapter, which contributes greatly to the spread of weeds by man.

4.3   Good powers of vegetative reproduction and ability to regenerate when divided into fragments

Many of the most serious and intractable weeds both in Britain and in other parts of the world are perennials which show vigorous powers of spread and reproduction by vegetative means. Bracken, couch, field bindweed, and ground elder are familiar examples of such plants. The ability to reproduce vegetatively, however, does not necessarily mean that sexual reproduction is not important. From the evolutionary point of view of course, the genetic variation achieved by sexual reproduction is vital, but in many perennial weeds the actual process of colonization and spread is greatly augmented by means of seed. Table 7 summarizes some of the characteristics of a small number of important and familiar perennial weeds, and it will be noted from this that seed production is important in several species though it may be unimportant (or indeed absent altogether) in others. One of the major disadvantages of a perennial habit which is solely dependent on vegetative means for spread is that really intensive cultivation which removes or kills all the fragments of the plant left in the soil leaves nothing to cause re-infestation. In the case of plants which have their organs of vegetative reproduction near the surface, adequate seed production to compensate for this type of eradication may be crucial. It is noteworthy that ground-elder, which has shallow rhizomes and in which seed production is insignificant is not a serious weed in agriculture, though it is in gardens; common couch, also with shallow rhizomes but which has adequate seed production, is an important weed in both situations.

Table 7 Some characteristics of a few important perennial weeds. (Based on ROBERTS, CHANCELLOR and THURSTON, 1968.)

Species	Reproductive parts: over-wintering state	Average depth of vegetatively reproductive parts*	Seed production
Aegopodium podagraria (ground-elder)	Rhizomes; dormant buds underground overwinter	Shallow (0–25 cm)	Unimportant
Agropyron repens (common couch)	Rhizomes; dormant buds underground but aerial shoots overwinter	Shallow	Fairly important
Allium vineale (wild onion)	Offset bulbs and bulbils; both overwinter	Aerial or very shallow (0·5 cm)	Rarely produced
Armoracia rusticana (horse-radish)	Taproot; this overwinters	Deep (> 25 cm)	None
Cardaria draba (hoary cress)	Creeping roots; small rosettes of leaves overwinter	Deep	Important
Convolvulus arvenis (field bindweed)	Creeping roots; these overwinter	Very deep (down to > 3 m)	Very important; only set in warm summer
Oxalis spp.	Bulbils, taproots, rhizomes; all overwinter	Shallow	None in some spp. Important in a few
Ranunculus repens (creeping butter-cup)	Procumbent creeping stems; some leaves overwinter	Above ground	Very important
Rumex crispus and *R. obtusifolius* (curled and broad-leaved dock)	Tap roots; rosette of leaves overwinters	Very shallow	Very important

* This may vary greatly according to soil type and other factors.

The major advantage of the perennial habit in weeds is that reserves of food are available both to support vigorous growth at the start of the season and also to sustain the plant through periods of heavy competition from crops. The major short-term disadvantage is the susceptibility to control by cultivation, but the possible limitations in the dispersal of propagules and the genetic uniformity imposed by a lack of seed production in some cases are also factors militating against the success of this mode of existence. Those plants which have survived as weeds in spite of these problems are formidable adversaries indeed for those concerned with weed control. For example, in the case of field bindweed one published recommendation for control was that the soil should be cultivated to a depth of four inches once a fortnight for three years! Many perennial weeds may be controlled adequately without such draconian measures. Their relative sensitivity to physical disturbance explains why most perennial weeds are a less serious problem in land cultivated for arable crops than in other types. In Britain *Agropyron repens*, *Agrostis gigantea* (black bent) and *Cirsium arvense* are found in arable land, but many other perennials are confined to small scale units such as gardens, smallholdings and other horticultural enterprises where a diversity of habitats, some relatively undisturbed, is available.

In pastures the predominant weeds are perennial, with such species as the docks, dandelions, plantains, yarrow, wild onions and common ragwort being frequently found. The last is normally a biennial, but may also be a short-lived perennial. The precise composition of the weed flora of pasture depends on the use to which the land is being put. The taller, ranker weeds are able to survive in long grass being grown for hay or silage, whilst shorter plants such as the daisy and the various species of hawkbit (*Leontodon*) are normally found where the grass is kept shorter by grazing.

The stored food reserves in the perennating organs of perennial plants are naturally not always at a constant level, and this may be important in relation to control. In temperate climates there is normally a considerable burst of growth in the spring and early summer which depletes the reserves. They tend to build up again later in the season when the plant has an ample leaf area and when, in some cases, crop competition may be decreased. Cultivations which take place at the period of lowest food reserves are more effective in controlling such plants than those carried out very early or very late in the season when reserves are higher.

It is an obvious characteristic of perennial weeds (as of other perennials with similar growth habits) that fragments broken off from the main plant will tend to regenerate into new individuals. In any situation characterized by physical disturbance this is an important attribute, though in the less disturbed environment of many perennials separation of new individuals from the parent may take place by actual abscission, as in the case of bulbils, or by damage to or decay of part of the plant between the parent and the vegetative offspring. Such offspring, often known as *ramets* to distinguish them from those sexually produced, may already be physiologically independent of the parent at this stage, possessing both shoots and roots of their own (see for example NYAHOZA *et al.*, 1973). In this case severing of the link with the parent plant simply completes the separation. In many cases, however, what happens is that an underground portion of the plant is separated. This bears dormant buds (if a rhizome) or no buds (if a root). In the latter case rapid bud initiation occurs and shoots grow upwards using the food reserves in the severed fragment until photosynthesis commences. In the case of the rhizome, severance from the main plant or from the rhizome apex may cause one or more of the dormant lateral buds to grow out. When this happens it is usual for one such new shoot to establish dominance over the remainder (CHANCELLOR, 1974). Fig. 4–1 shows the pattern of growth of the shoots from buds on a seven node fragment of *Agropyron repens* rhizome over a period of twenty days.

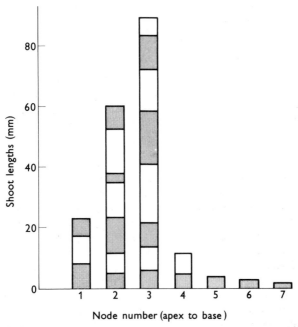

Node number (apex to base)

Fig. 4–1   Diagram of shoot growth from a seven node fragment of the rhizome of common couch over a period of 20 days. Horizontal lines on bars represent (from the bottom) lengths after 3, 5, 7, 10, 12, 14, 17, and 20 days respectively. (Redrawn from CHANCELLOR, 1974.)

Fig. 4–2 shows the pattern of growth in the soil of a single plant of field bindweed. It is clear from this that cultivation to different depths would leave differing numbers of shoots and roots still attached to the principal root system, and regeneration from these would be rapid and vigorous. A striking and unexpected feature in this plant, which is discussed again in chapter 6, is that isolated fragments of roots rarely give rise to new plants, since there appears to be a failure of root formation in such fragments although new shoots appear very quickly. (Chancellor, private communication; unpublished data of Ismail and Hill.)

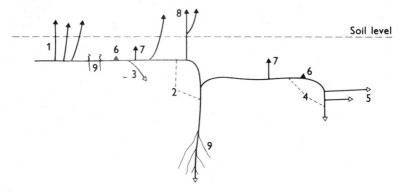

Fig. 4–2 Schematic representation of the extent of lateral spread of the roots of field bindweed in a non-competitive situation. (1) Parent plant. (2) Primary lateral root. (3) Lateral root branch. (4) Secondary lateral root. (5) Tertiary lateral roots. (6) Buds more than 5 mm long. (7) Developing shoot. (8) Aerial shoots. (9) Short feeding roots. (After DAVISON, 1970.)

Another factor which may be important in the survival of rhizomatous plants under cultivation is the depth to which rhizomes penetrate and from which fragments can regenerate. Soil factors may markedly affect both the rate of spread and depth of penetrations. Fig. 4–3 shows rhizomes of the hedge bindweed (*Calystegia sepium*) growing in a deep, narrow box filled with John Innes compost. In this light medium, rhizomes have penetrated to a depth of about 60 cm though in the field they are usually confined to the upper 30–35 cm of the soil. In this plant even 3 cm single node fragments showed the ability to produce shoots which reached the surface from a depth of 40–50 cm, though the percentage able to do this was low. 100% of such fragments regenerated new plants from a depth of 10 cm (unpublished experiments of Ismail and Hill). Common couch rhizomes rarely regenerate if fragments are buried more than 30 cm deep.

Perennials show seasonal patterns of growth just as do annuals, and there are periods of the year when buds on rhizomes may be effectively dormant or when regeneration from roots may be low. Little is known of the factors affecting this type of dormancy, and indeed many aspects of the physiology of rhizomes and perennating roots need more thorough investigation. Just as in the case of seeds, dormancy in vegetative parts can present great problems in control, since it enables propagules to survive cultivation. Conspicuous examples of the importance of this are found in the two species of nutsedge (*Cyperus rotundus* and *C. esculentus*) which are

Fig. 4–3   Rhizomes of the hedge bindweed (*Calystegia sepium*) growing in an easily penetrable soil (John Innes compost) in a container with a sectional door enabling part or all of the soil profile to be observed. The deepest visible rhizome has penetrated to a depth of about 60 cm.

very serious weeds in warmer parts of the world. These possess stem tubers which show variable degrees of dormancy.

Another feature of perennial weeds which should be noted here, though it is dealt with in more detail in chapter 5, is that the proportion of them which are polyploid tends to be greater than that in the general population of perennials. In view of the fact that polyploidy is often (though not invariably) associated with vigour this feature may be an important one in some species.

4·4 Self-compatibility

Any plant which is able to set seed in the absence of either other plants of the same species or of appropriate pollinating insects is likely to be at an advantage in situations where colonization is taking place. It is therefore not surprising to find that a very large proportion of weeds are self-compatible. This matter is also developed more fully in the next chapter, though it has an obvious place in any list of desirable weedy characteristics. Associated with it as a characteristic related to pollination is the fact that where cross-pollination is at all common, it is clearly advantageous to weeds if the pollinating medium is either wind or insects with non-specialized habits. Both situations are in fact common in weedy species, though this is not to suggest that they are uncommon in others.

4·5 Strong competitive ability

We have already discussed in some detail the ways in which weeds compete with crops. It is apparent that they do not all do this in the same way, but it seems obvious that 'competitive ability', whatever this is, must be a vital characteristic of any successful weed. The component elements of competitive ability are varied. They include such physiological attributes as efficient ion uptake and rapid root growth, together with purely morphological or behavioural features such as the ability to climb up other plants (bindweeds), to scramble over competitors, thus greatly reducing the light available to these (common chickweed), or to produce rosettes of leaves which spread out close to the ground and smother competitors in this way (daisy, plantains, dandelion). Sheer size in itself is not by any means a necessary component of competitive ability, though some weeds are conspicuously large and vigorous (e.g. charlock in cereals). It is not the possession of such features individually which makes a weed a good competitor, but the combination of them in a particular plant which is thereby suited to some special position in the spectrum of weeds which may affect crops. When we talk of measuring competitive ability what we usually mean is the measuring of some arbitrary *result* of such ability, such as reproductive potential or vegetative spread (see also the end of chapter 5).

5 Weed Spread and Weed Evolution

5.1 The spread of weeds by man

Within any particular geographical area individual weed species are often widespread, and a considerable number of weeds are virtually cosmopolitan. Some familiar ones in this category are bracken, chickweed, knotgrass, annual meadow grass and charlock. Some natural adaptations of weeds to dispersal have been discussed briefly in chapter 4, but one important characteristic of weed dispersal is that it is to a very large extent due to man and his activities. The passage of weedy species over most major geographical barriers such as mountain ranges and oceans is nearly always a result of human agencies.

Happily for weed biologists some examples of the spread of weeds are very well documented. In a small country like Great Britain, where some of our important weeds have a long history, this is not always easy, but in the United States where a very large proportion of the weed flora has been introduced, the spread of many species can be traced with considerable certainty. For example, it is known that most of the weeds in the flora of North America have invaded from Europe and have spread across the country from east to west along with successive waves of pioneers. It is because of this that ribwort plantain (*Plantago lanceolata*) is known to the American Indians as 'White Man's Foot'. Relatively few have spread from west to east or in a north–south direction. It is also interesting that although many native European plants have become successful weeds in the New World there are very few examples of New World plants which have returned the compliment. Amongst those which have done so are *Conyza canadensis* (Canadian fleabane), *Helianthus annuus* (sunflower) and *Solidago canadensis* (Canadian golden rod), which often occurs as a garden escape.

It is clearly of interest to look in a little more detail at some of the ways in which weeds may be spread by man. One excellent reason for doing so is that in some cases it is possible to take precautions to minimize the risk of spreading weeds, though when the range of methods of spread is considered it is obvious that we stand very little chance of entirely preventing any weed from spreading to a new area in this way.

Table 8 summarizes some of the associations between man and weed spread, and these are then discussed further.

Of the methods listed one of the most important is probably the introduction of weed seeds as contaminants with crop seed. There are a number of reasons for this. One is that such contamination is virtually inevitable. However vigorous the cleaning procedures may be, none is perfect when large amounts of seed are being handled, and consequently

Table 8   Some ways in which weeds may be spread through human agency.

1   Contaminants in seed crops
2   Accidental spread of waste from cleaning of crop seed
3   Contaminants in hay, silage and other animal feeding stuffs
4   Contaminants in straw
5   Animals: (a) internally, (b) externally
6   Packaging materials
7   Machinery in normal cultivation
8   Bulk movement of sand, ballast and soil
9   Deliberate introduction as garden plants, herbs, curiosities, etc.

weeds are almost always sown with crops. A second and almost equally important point is that the weeds thus sown are likely to be those which have already proved successful competitors with the particular crop and suited to the environment in which the crop grows. When one considers the fact that the weed seeds will be sown well spread out with the crop, and may consequently have the opportunity to establish over a fairly wide area, it becomes clear that this is a very important method of spread. Tables 9 and 10 show data on the presence of weed seeds in samples of

Table 9   The occurrence of certain weed seeds in cereal samples. Figures represent the percentage of samples with at least one seed in 4 oz. (113 g). (From data of WELLINGTON, 1960.)

Weed species	Wheat	Crop sample Barley	Oats
Galium aparine	8	12	12
Polygonum convolvulus	6	17	11
Polygonum aviculare	3	4	4
Rumex crispus	2	4	9
Avena fatua	2	7	2

Table 10   The occurrence of certain weed seeds in samples of red clover and lucerne. Figures represent the percentage of samples containing at least one seed in 5 g. (From data of WELLINGTON, 1960.)

Weed species	Crop sample Red clover	Lucerne
Plantago lanceolata	35	56
Rumex crispus	26	19
Lychnis spp. (ragged robin, catchfly)	17	9
Geranium dissectum	16	1
Plantago major	9	2

crop seeds. HARPER (1960) pointed out that a seed sample of a cereal sown at a rate of 1.5 cwt/acre (188 kg/ha) would, if contaminated to the extent of 1% with the relevant seeds, give a sowing density of 5 seeds per square foot of cleavers, 2 per square foot of wild radish and 2–3 per square foot of knotgrass. None of these represents a high initial density, but if established they could lead to a substantial build-up of seed in the soil, which would be reflected in much heavier infestations in the following year.

Where crop seeds are cleaned prior to sale there is an obvious danger from the accidental spread of the cleanings. In this case, however, precautions against seed spread can be made reasonably effective. In Great Britain and in many other countries there are stringent regulations relating to the maximum permissible contamination by weed seeds in commercial seed samples, and one of the reasons for legislation on the import of certain kinds of seeds into some countries is the risk of introduction of weed contaminants. In addition many crops which are grown to provide seed have to be inspected before harvest, and are not acceptable for the purpose if specified densities of certain weed species are exceeded.

Farm animals are an important cause of the spread of seeds in a variety of ways. Since they are moved with and by man the fact that many weed seeds will survive passage through their alimentary tracts can be a source of weeds in new areas. The same applies to those weeds whose seeds become entangled in hair or wool. There are very many records of seeds being found in the cleanings from wool and in the 'shoddy' which is a by-product of wool processing and which was at one time widely used as a fertilizer (see SALISBURY, 1961, pp. 138–42). Clearly a very large proportion of the alien species introduced by this method will have only a transient life in a new country, but the possibility of the spread of weeds by such means is certainly a very real one.

In addition to direct spread by animals there is also the fact that weed seeds may be spread in animal food such as hay and silage. In the latter both the temperatures reached and the chemicals produced during the process of making it may reduce the viable weed seed population in some parts of the silo, but not in others. Straw for animal bedding is another possible source of weeds, and weeds have certainly been spread by a combination of these means in many cases. For example the devastating infestations of perforate St John's wort in northern California which took place in this century initially followed the cattle and sheep trails which were used in the area.

What has already been said about the weed seed population of soils in chapter 3 makes it clear that any agency by which soil is moved from place to place may be an important factor in the movement of weeds. Agricultural machinery is an immediate candidate in this connection, for though weeds are not likely to spread over very long distances in mud attached to machinery, short distance transfer of weeds from one location to another by this means is probably frequent. Not only annual weeds but also perennials may be spread in this way. The same naturally applies to the bulk movements of soil, ballast, sand and other materials which may take place in connection with building operations. In large-scale enterprises such as road-building the chance of the spread of local populations of a weed species to other relatively remote areas in this way must be considerable.

Packing materials form yet another medium for the spread of weeds. Whenever straw or other plant products are used in packaging this possibility exists, but it may be particularly important in the transfer of weeds with nursery stock in the horticultural industry. Whenever whole plants are moved they tend to be packed with materials, including soil, which may form a source of weed infestation. CRAFTS and ROBBINS (1962) point out that camel's thorn (*Alhagi camelorum*) was introduced into California from the Middle East in the packing round date palm shoots, and give other examples of the same type of introduction. The introduction of pernicious perennial weeds by this means is a particularly serious possibility.

The fact that some weeds have seeds well adapted to dispersal by water has also been mentioned earlier, but it is worth pointing out that many plants have seeds which can survive periods of days or weeks in water without undue loss of viability. The spread of weeds along irrigation or drainage channels by this means is certainly common, and it has been shown that enormous numbers of seeds may be moved by water in such channels. The further such water is moved the more important this method may be, but the significance of relatively local shifts of infestation should once again be pointed out.

A final method of weed spread which must be mentioned is deliberate introduction by man. Many weeds were at one time used for their medicinal properties and were probably introduced for this purpose as man moved from one area to another. In another connection the distinguished weed biologist Harper, in a comment in discussion of a paper by BAKER (1965), said: '. . . it seems to me worthwhile to point out how potentially dangerous botanic gardens are.' He added that the catastrophic effects of the water fern *Salvinia* in rice fields in Sri Lanka (Ceylon) have followed its introduction from a botanic garden. The same point might be made about plants grown in ordinary gardens. *Oxalis pes-caprae* is a very serious weed in parts of Australia and in the bulb fields of the Scilly Isles, and was introduced to both as a garden plant. Many other examples are quoted by SALISBURY (1961). The problem here of course is that it is never possible to judge whether a plant is likely to turn into a weed when it is introduced into a new area. Perhaps one of the roles of the study of weed biology is to draw attention both to the potential of the problem and also specifically to those characteristics of plants which might be important in making them successful as weeds. For example, perhaps any plant showing vigorous vegetative reproduction or a high reproductive potential in general should be regarded with caution.

5.2   The success of aliens as weeds

Before coming to a discussion of some aspects of the evolution of weeds it may be useful at this point to consider why it is that so many plants have become weeds only when transferred from their native to an alien environment. Several possibilities suggest themselves immediately as likely to be very important.

(a) A plant transported to a new area may well be removed from the influence of pests and diseases which keep it in check in its native environment. Very good examples of this provided by the prickly pear (*Opuntia* spp.) in Australia and by perforate St John's wort in California. In both cases the introduction of the appropriate pest achieved spectacular control in a very short time. It is very relevant to note that this general point applies also to crops. A large proportion of the world's crops have their main centres of production far from the areas where their ancestors are or were indigenous, and an important factor in this is the absence in new areas of pests and diseases prevalent in the old.

(b) A plant may by chance be better suited to the particular environmental conditions in the new environment than it is to its native environment. This possibility is clearly related to the next.

(c) In the new environment there may be an absence of particular constraints which operate on the plant in its home territory. The natural flora of any area contains those plants which, by natural selection, have become adapted to specific ecological niches. It is always likely that in any environment there may be niches which have not been filled, and an alien species may thus be able to take advantage of a type of 'ecological vacuum' in which the competitive pressures on it are different from and may be less than those in its native environment. If the physical environment also happens to favour the growth of the plant more than its native one these two factors will combine to produce a plant with quite different ecological characteristics in the new situation from those it showed in the old. Once it is established, of course, the ordinary processes of natural selection will begin to operate on populations of the alien species, and weedy variants may be selected which could not have emerged under the different selection pressures of the native environment. The really critical phases in the establishment of an alien in a new environment are likely to be those of seed germination, seedling establishment and the interaction of the young developing plant with the indigenous flora. Any alien which can reach maturity and set seed is a possible forerunner of a race of weeds, but in order to keep this possibility in proportion it should be re-emphasized that the total number of weed species in the world is really remarkably small in relation to the world flora. Only a tiny fraction of the plants whose seeds or other propagules are transported to new environments each year have more than an extremely transient existence as aliens; of those which do very few indeed are likely to become weed problems. One might speculate on whether the evolution and spread of weeds has already been so successful that the conditions imposed on those new arrivals which are putative weeds have become too demanding for the evolution of many new species of weed to be at all likely.

5.3 The evolution of weeds

What follows is not in any sense intended as an exhaustive account of weed evolution. It simply points towards a few of the interesting problems which have been considered in relation to the subject and which seem to throw light on the overall theme of the book, which is that of weeds as a biological phenomenon worthy of study in their own right, as well as in relation to their control.

We have already seen that weeds are likely to have originated firstly from plants of open and disturbed habitats present before the advent of man, and secondly, from plants which became associated with man as soon as he started to cultivate, and indeed earlier. It is a characteristic of plants which live in very demanding environments, where the selection pressure against poorly adapted individuals is very strong, that they show very clear and rigid adaptations to their environment. Desert plants and those of high mountains provide good examples of this. Now the selection pressures operating in disturbed environments are also very strong. They may include considerable extremes of temperature, nutrient shortage, susceptibility of the soil surface to compaction and other features. When these natural hazards are supplemented by the vigorous pressures due to the agriculturalist (whose cultivations are designed to a considerable extent to combat weeds) the evolutionary pressures are seen to be substantial indeed. Successful weeds, as we have seen, share many characteristics, though perhaps the only one common to every species is the ability to survive in cultivated land. Rigidity of adaptation to fine variations in the environment is also a prime characteristic, and this is one of the features which needs to be thought about in considering the evolution of weeds.

In view of the relatively small number of weed species in the world it is not perhaps surprising to learn that few genera contain more than a few weedy species. Amongst those in the British flora which do are *Polygonum* with seven and *Fumaria* and *Veronica* with ten each. These facts can be added to that noted in chapter 1 that many of the most important weeds belong to a relatively few plant families. Most of these are evolutionarily advanced families, with a high proportion of herbaceous members. There are, of course, a number of woody weeds; some of the plants which are weeds of range lands fall into this category. Foresters recognize certain shrubs and trees as weeds in their particular environment. *Calluna vulgaris* in spruce plantations and, formerly, silver birch and hornbeam in various situations fall into this category. In spite of this most weeds are annual or perennial herbaceous plants. The biennial habit does not appear to lend itself to the development of weedy species, though there are exceptions such as ragwort and corn parsley (*Petroselinum segetum*) together with some other weeds which may behave either as annuals or as biennials. In Great Britain there are, however, a number of very successful biennials which are weeds of roadsides and waste places where the pressures due to cultivation are not so great.

From the evolutionary point of view there is no necessary difference between annuals and perennials except in their way of resisting unfavourable seasons, but where reproduction in perennials is largely or entirely by vegetative means rather than by seed there are disadvantages to the species. Resistance to assiduous cultivation may be much reduced (e.g. ground-elder); dispersal over longer distances is less likely, and most importantly the genetic variation achieved by sexual reproduction is reduced or lost altogether. Those plants which have been successful in spite of some or all of these disadvantages form some of the most successful weeds of all, and include such plants as common couch, creeping thistle, Johnson grass (*Sorghum halepense*), the nutgrasses (*Cyperus rotundus* and *C. esculentus*) and many others.

One advantage of the perennial habit is that the reproductive capacity of some plants is reduced when they are near the limits of their geographical range and the ability to reproduce by vegetative means can compensate for this.

It has often been suggested that weeds show a greater tendency than other types of plants towards polyploidy. This is the condition in which a plant contains in each cell a number of chromosomes which is some multiple of the basic number for the group. Polyploids arise in two principal ways. In the first the chromosome number of an individual is doubled by chromosome division without cell division at a critical stage of development, leading to the production of a plant in which every cell contains not two but four sets of homologous chromosomes. This is known as autopolyploidy. In the second case this chromosome doubling is preceded by the formation of a hybrid between two different species. These may be sufficiently distantly related that the hybrid is sterile because chromosomes do not have homologous partners with which to pair in meiosis. Doubling of the chromosome number in such hybrids may restore fertility by giving each chromosome an exact homologue. This is termed allopolyploidy. Polyploidy is often, but not invariably, associated with an increase in size and vigour in plants and the idea that many weeds may be of polyploid origin may have been initially suggested by this. In fact it appears that in many cases, certainly in annuals, the proportion of polyploids does not exceed the general proportion in the natural flora, though in perennial weeds it may do so, especially in those which show vigorous vegetative reproduction. For example, two thirds of the perennial members of the Swiss weed flora have been shown to be of polyploid origin, whilst the same is true of only one third of the annuals. In Canada the proportion of polyploids amongst the most common weeds does not differ from the rest of the flora, though in perennials the proportion of polyploids in weeds of certain specialized habitats was rather higher (MULLIGAN, 1965). The association of polyploidy and weediness needs further study, but it is not surprising to find that any character conferring even a marginal tendency towards vigour, and hence perhaps competitive ability, may have been selected for in weedy species.

Another important characteristic of weeds in relation to their evolution is their tendency to be self-pollinated. This has already been referred to in chapter 4. Some weeds are invariably self-pollinated (e.g. shepherd's purse); a very large proportion are normally self-pollinated but also show a variable proportion of cross-pollination, whilst a few are actually self-sterile. Examples of this last group are the weedy poppies and sunflowers. An even smaller number are dioecious, having the two sexes on separate plants. A good example of this is the creeping thistle, which compensates for the obvious disadvantages of the dioecious habit by having formidable powers of vegetative spread and reproduction. The advantages of self-fertility and a large percentage of self-pollination is that seed is produced even when a plant is not one of a reasonably dense population. This could be extremely important in the initial establishment of alien species. The disadvantages of self-pollination are that if it is continued for a few generations it results in almost one hundred per cent of the population being homozygous for all characters; this leads to a lack of variability in individual self-pollinated lines since offspring will come to be genetically identical to their parents. In this situation any difference between individuals is due either to a difference in parentage or to the environment. In obligatory self-pollinators such as shepherd's purse great differences in the morphology of individuals may be found due to both these factors, but for those characters which are genetically determined each plant will breed true. In a situation where the environment may change, such genetical invariability is, or can be, a serious drawback. The critical importance to weeds of plasticity in relation to the environment (what geneticists would call phenotypic plasticity) becomes obvious in such situations.

A compromise between the genetical invariability of populations derived by self-pollination and the variability of those where cross-pollination is the rule can be achieved if cross-pollination occasionally takes place. This can lead to the maintenance of a certain amount of heterozygosity, and hence variability, in the population without the sacrifice of the advantages of self-pollination. BAKER (1965) suggested that what weeds required was a 'general purpose genotype'. This is a genotype which allows a wide degree of phenotypic plasticity and an adequate and sustained level of heterozygosity. One way in which heterozygosity may be maintained is, of course, by vegetative multiplication and in the case of the Bermuda buttercup (*Oxalis pes-caprae*) this is certainly what happens. Another way of 'fixing' heterozygosity where this contributes to plasticity is by apomixis, where zygotes are formed without fertilization of the egg by the gamete from the pollen grain. This phenomenon is sometimes found in weeds, good examples being the dandelion and perforate St John's wort. In the former it is invariable, whilst the latter is able to reproduce apomictically but does not always do so. In the British weed flora apomixis is not in fact particularly common especially amongst annuals, but there is no doubt that it can be an important factor in the evolution of species in which it does occur. Heterozygosity may also be conserved in allopolyploids where there is duplication of many genes due to the presence of four (or more) homologous chromosomes.

There are other genetic phenomena which can permit the maintenance of heterozygosity in spite of a very high proportion of self-pollination, but the details of these need not concern us here (see, however, BAKER, 1965; EHRENDORFER, 1965). It is sufficient to say that all such mechanisms appear to be common in weeds and would have led during evolution to the maintenance of genotypes allowing a very high degree of phenotypic plasticity without the loss of ability to evolve new forms through natural selection.

Interesting comparisons can be made between weedy and non-weedy species in the same genus, and these throw light on some of the factors which may have been significant in the origin of the weedy forms. Table 11 shows a comparison between species of the genera *Eupatorium* and *Ageratum,* both members of the Compositae. There are two main points of interest which emerge from this table. One is the extremely close parallel between the two genera in the differences which distinguish the weedy from the non-weedy species. The second is that at the chromosomal level it is obvious that these differences have evolved along quite distinct paths. *Ageratum conyzoides* appears to be a polyploid relative of *A. microcarpum,* whilst the weedy *Eupatorium microstemon* has only one fifth of the chromosome complement of its non-weedy relative. Mechanisms leading to this type of reduction in chromosome number are well known (see STEBBINS, 1971) and are amongst those by which heterozygosity may be maintained.

Another study of weedy and non-weedy forms, this time of the same species, perforate St John's wort, is that of PRITCHARD (1960). This is a species which in Europe shows little tendency to weediness, whilst in other parts of the world it has become a serious weed. When specimens from four different areas were grown together under the same conditions they fell clearly into four groups on the basis of height. Tallest were the plants which were Australian weeds; next were more weedy British forms, followed by a woodland form, with the shortest being a sand dune type. This suggests that, in this species, ecotypes have been developed in the different environments which have fitted the plants for their particular mode of life. This phenomenon is very well known in plants generally, but is interestingly less usual in weeds than in other types of plant. It has been

Table 11   A comparison of the characters of closely related weedy and non-weedy forms in two genera. (Modified, with permission, from BAKER, 1965.)

1	*Eupatorium*	
	E. microstemon (weed)	*E. pycnocephalum* (non-weed)
	(n=4)	(n=20)
	Plastic	Not very plastic
	Annual	Perennial
	Quick to flower	Slow to flower
	Photoperiodically neutral	Short-day requirements
	Self-compatible	Self-incompatible
	Economical pollen production	Plentiful pollen production
	Tolerant of drought and waterlogging	Intolerant of drought and waterlogging
2	*Ageratum*	
	A. conyzoides (weed)	*A. microcarpum* (non-weed)
	(n=20)	(n=10)
	Plastic	Not very plastic
	Annual	Perennial
	Quick to flower (6–8 weeks)	Slow to flower (flowers in second year)
	Flowers with low or high night temperatures	Flowers with rather low night temperatures
	Flowers in any daylength	Flowers better in short days
	Self-compatible	Self-incompatible
	Economical pollen production	Plentiful pollen production
	Tolerant of drought and waterlogging	Intolerant of drought and waterlogging

suggested that this is because the plasticity which weeds show in relation to one environment tends in one sense to remove the need for the formation of genetically stable ecotypes. What is needed in weeds is not a fine morphological and physiological adjustment to a specialized habitat but a very flexible adjustment to a highly variable one. A tendency to form pure breeding ecotypes would conflict with Baker's ideal 'general purpose genotype' and could be a positive disadvantage to a weedy species. A response to environmental stress through plasticity is likely to enable a range of genotypes to flourish in the same population, whereas a response to stress which led to the rapid elimination of unsuccessful and non-flexible forms would reduce the genetic diversity which may be important in the general evolutionary history of weeds. It is interesting that one type of open habitat which does not appear to have produced many weedy species is the desert. A probable reason for this is that survival in these particularly harsh conditions really does require very close adjustment to the environment at the physiological level, and that the 'general purpose genotype' is unsuited to this in itself, and is less likely to evolve from plants successful in this environment.

In chapters 2 and 4 there was much reference to the rather general character 'competitive ability' which it seems certain that weeds must possess for success. If this is so then it must be assumed that natural selection will have perpetuated those weeds showing the character. It is worth noticing that, despite the very great difficulties of defining and measuring competitive ability in a meaningful way, there have been studies which have not only done this but have attempted to assess whether it is a character which can be inherited (e.g. SAKAI, 1961). These studies have shown that on the basis of the criteria of measurement used competitive ability is such a character. It can be inherited in the same way as many other so-called quantitative characters in plants, such as yield and growth rate. Such characters are controlled by many genes acting together, and although each gene behaves in a strictly Mendelian manner the precise effect of each gene is obscured by the others which all affect the same process. The statistical and genetical niceties of the argument here are not crucial to the main point, which is that the ability to compete successfully can be selected for and inherited. It cannot be denied that, whatever the criteria of measurement, weed evolution has been extremely successful in producing plants adapted in a whole complex of ways to their highly characteristic mode of life.

5.4   The future of weed biology

There are more people working on various aspects of the biology of weeds at the present time than ever before. A very large proportion of these are doing work very directly related to the control of weeds, but there are still many studies going on which emphasize the concept of understanding the plant and its way of life as a prelude or a key to the development of control measures. SAGAR (1968) in a stimulating, if perhaps over-optimistic speculative paper on the future of weed biology, made a number of highly pertinent comments. He emphasized the great need for communication and collaboration between workers in different disciplines with a common interest in weeds. There is certainly a surprising shortage of critical physiological work in some areas of weed biology and interesting problems abound. It may well be true, as Sagar suggests, that the number of cases where detailed study of the biology of the weed has actually contributed positively to the development of control measures may be small. Control measures adopted by farmers have themselves evolved by a process of natural selection, since farmers cannot wait for a weed biologist to study their problem before they try to do something about it. That there have been some successes in this field is perhaps a bonus, since the study of weeds as a phenomenon has certainly contributed in many ways to our understanding of many aspects of biology, and will continue to do so. Where highly sophisticated herbicides are available, enabling us to control a wide range of weeds at will, the role of the weed biologist as such may come to appear peripheral to those of the biochemist and agricultural engineer as a contributor to the technology of weed control. It will none the less be a long time before this situation applies on a worldwide scale, and in underdeveloped areas and those where agriculture is spreading, weed biology still has much to contribute.

6 Experiments with Weeds

It is probable that some of the earlier parts of this book, especially chapters 2, 3 and 4, may give anyone interested many ideas for experiments which might be possible with weeds. One advantage of weeds is that there is rarely if ever any real shortage of experimental material, though in some cases a certain amount of forward planning may be needed if, for example, seeds or seedlings are to be used.

The suggestions for experiments made here are simply pointers—ideas on which a large number of variations are undoubtedly possible, and what is certainly true is that carrying out any one of them will give an experimenter ideas for others.

A problem which does arise when actual experiments with weeds are contemplated is that of identification of plants. Many of the plants which may be involved will be fairly well known, if not to the experimenter at least to any friend or relation who happens to be a gardener or farmer. A large proportion of the plants which make up the normal weed flora can be identified readily enough with the aid of a flora such as CLAPHAM, TUTIN and WARBURG (1968). A particular problem in weed studies is that it may be necessary or useful to identify quite young seedlings. This is the case, for example, in the experiments described on weed germination from soils of different areas. Happily the excellent book by CHANCELLOR (1966) makes this a relatively easy matter. This book contains keys for weed identification based entirely on the characters visible in the very young seedling, and is illustrated by line diagrams of an elegant simplicity which is extremely helpful. Figure 6–1 shows three species as illustrated by Chancellor as examples.

Another problem which can beset experiments with weeds is that of seed germination. In experiments designed to look at the phenomenon, the variable nature of weed seed germination is simply a matter of interest. If, however, one is concerned to obtain a large supply of uniform seedlings for some other purpose very real difficulty may be experienced. Preliminary tests on the germination rate of seeds may be necessary to establish the number which will be needed in order to provide the required number or density of seedlings. This applies, for example, to experiments where a crop and a weed are to be allowed to grow together in order to observe their mutual interference.

Some of the experiments described can be done quite simply in the laboratory, on the bench, on windowledges or in incubators. Others really need a certain amount of space in a glasshouse or conservatory or perhaps in a small area of a garden. Anyone trying to carry out experiments with weeds in a garden, especially somebody else's garden, may need to use a certain amount of tact in the preparatory stages in order to reassure other users, and particular care should be taken in these circumstances if 'difficult' perennial weeds are involved.

A last problem in connection with these experiments which should be mentioned is that, as in all experiments with living plants, and especially with highly plastic and variable plants such as weeds, results are likely to show wide variations. Some care in the design of the experiment can minimize the effects of this variation, and the books by HEATH (1970) and

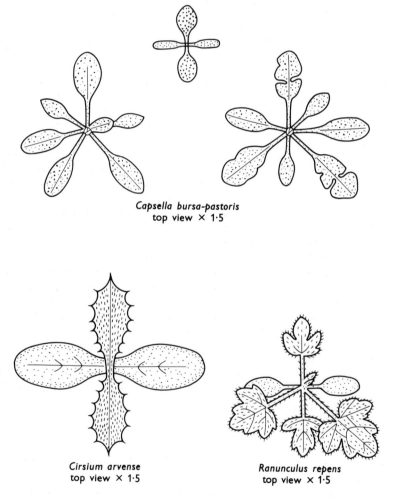

Capsella bursa-pastoris
top view × 1·5

Cirsium arvense
top view × 1·5

Ranunculus repens
top view × 1·5

Fig. 6–1 Seedlings of three common weeds as illustrated by CHANCELLOR (1966).

PARKER (1973) offer detailed guidance on statistical treatment of experimental data. Even with plants as variable as weeds, however, it will often be possible to obtain results which are readily interpretable without statistical methods being applied. A safe rule of thumb might be that if a result is not reasonably clear without statistical help then the experiment really needs to be repeated in any case.

6.1   Some experiments with perennial weeds

(a)   Regeneration from perennating organs

Many experiments are possible on aspects of the problem of regeneration of plants from the perennating organs of perennial weeds (often roots, stolons or rhizomes). It may be of interest to determine the effect of the size of fragment of such organs on the speed and vigour of regeneration, since this may have some importance in relation to cultivation procedures. The 'physiological age' of the fragment may be important also and this can be examined by comparing fragments cut at different distances from the apex. (The morphological apex may be easier to find in underground stems than in roots.) Experiments of these kinds may be done in pots or shallow trays of soil, sand, peat or other suitable medium. It will quickly become apparent in such experiments that one of the most important things to decide is what to record and how and when to record the results. For example, those fragments which regenerate most rapidly will not necessarily produce the largest or most vigorous plants. In addition root fragments may develop more than one shoot, whilst rhizome fragments will normally develop only one shoot for each bud, and where fragments containing more than one bud are used there may be a relationship between the behaviour of one bud and that of the remainder (CHANCELLOR, 1974). Such things as the number or percentage of fragments regenerating, the time taken to reach some particular stage of growth (or the stage of growth reached after some fixed time) may be all usefully recorded for different purposes.

The effects of various environmental factors on the regeneration of plants from fragments may also be looked at, and some of these can be considered with very simple facilities indeed. Some suitable plants for such experiments are common couch, ground-elder, hedge bindweed, creeping thistle and, subject to the problem mentioned later (section *d*), field bindweed.

(b)   The effect of depth of burial on regeneration from fragments of perennating organs

This factor is obviously important in the field, and a simple way to study it is illustrated in Fig. 6–2.

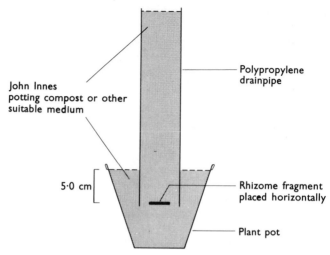

Fig. 6–2 Arrangement for investigating the regeneration of plants from rhizome fragments buried at different depths.

The use of lengths of polypropylene drainpipe inserted in plant pots makes it possible easily to plant rhizome or root fragments (or indeed seeds) at varying depths in containers of the same size, and samples can be looked at one at a time without disturbing the remainder, which is valuable in recording results.

(c) Effects of some soil conditions on regeneration

Regeneration of plants from fragments of weed roots or rhizomes in different types of soil can be examined, but handling some field soils, especially heavy ones, in the laboratory, can be difficult and problems arise over watering and drainage. It is, however, easy to devise experiments to compare the resistance of fragments of perennating organs to drought or to short periods (up to say 10–14 days) of waterlogging.

(d) An experiment with field bindweed—a cautionary tale

Field bindweed has thick roots which serve for storage and for spread of the plant in the field. It is usually assumed that fragments of the root of this plant will give rise to new plants in the normal way. Experiments in our laboratory (Ismail and Hill, unpublished data) and elsewhere have shown that although such fragments rapidly give rise to new shoots these tend to collapse and die after a relatively short time (see Fig. 6–3).

This confirms some earlier reports that field bindweed regenerated much less well from root fragments than is often suspected. The

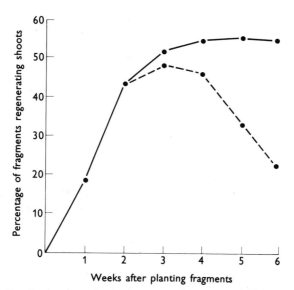

Fig. 6–3 Graph showing the percentage of 1 cm root fragments of field bindweed, giving shoot regeneration at various times after planting (· —— · —— ·) and the percentage whose shoots still appeared healthy when observed at the same times (· – – – · – – – ·).

phenomenon appears to be due to a failure of root production from the isolated fragments and would repay further study.

6.2   Some experiments with annual weeds

(a)   Introduction

Many annual weeds occur in large numbers in varying habitats in the same area. Some considerable interest may attach to a survey of one or more species to observe the variations in size, form, reproductive capacity and stage of growth at which flowering takes place in the field.

(b)   Experiments on young seedlings

Young seedlings may be transplanted from the field into pots containing different soils or into some uniform medium such as coarse sand, peat or vermiculite in which they can be treated differently in relation to the supply of nutrients, water, light, etc. The variations of form experimentally induced by this means can be compared with those observed in the field, and in this way some picture may be built up of the extent of environmental as opposed to genetical variability in the species chosen. Species suitable for experiments of this kind are those which are likely to be very common and frequent in many areas, such as annual meadow-grass, fat-hen, shepherd's purse, groundsel, red dead-nettle and several others.

In the study of variation in reproductive capacity in annual weeds considerable help and many ideas may be obtained from SALISBURY (1961). It is worth noting at the outset, however, that there are several components of reproductive capacity, all of which may vary, though some vary more than others. Some variables are, for example, the length of the flowering period, the number of inflorescences, the number of flowers in each inflorescence, the size of flowers, the number of fruits produced and retained, the number of seeds in each fruit (where relevant) and the size or weight of individual seeds. The last factor is frequently rather invariable in comparison with the others.

(c) Experiments on seed germination

These experiments do present problems because of the inherent variability in the germination of weed seeds, but on the other hand some germination experiments are relatively quick to carry out and require little in the way of space and facilities. Amongst many possibilities which suggest themselves for experiments on seeds actually collected from weeds in the field are the viability of seed collected before it is mature, the effect of age or storage conditions on germination, and the effect of environment on the process. Many weed seeds show a period of light-sensitivity in germination (see below) and some may germinate better if exposed, whilst damp, to low temperatures for a period of a few days or weeks. The necessary temperature of around 5°C is easily obtained in an ordinary domestic refrigerator.

In all germination experiments it will be found useful in the first instance to try more than one species, because some are much easier to handle than others. It is also useful to look up relevant details of what may be known of the periodicity of germination so as to avoid wasting time trying to germinate seeds at a time when a large proportion are dormant, unless this is a part of the experiment (see for example Fig. 1 in ROBERTS, CHANCELLOR and THURSTON, 1968). When germinating seeds on filter paper in petri dishes it may be useful to surface-sterilize them by washing for 2–3 minutes in the supernatant of a freshly prepared slurry of calcium hypochlorite (about 5% w/v), and then to wash them in sterile water before placing them in the dish. Notice, however, that the precise conditions of germination may be very important in some cases. SAGAR and HARPER (1960) showed that *Plantago lanceolata* germinated much better in soil than on filter pads even after prechilling for 14 days at 5°C, though the two other species with which they compared it did not show this effect.

(d) Experiments with weed seeds in soil

The importance of the weed seed population of the soil has been stressed earlier (chapter 3) and many investigations are possible to compare the weed flora of soils of different types, or of soils of the same type with different agricultural histories. Soil collected from different depths in carefully dug holes may be compared, and particularly interesting data may be obtainable by looking at soils from beneath pastures which have been present for varying periods. The sources of samples for such tests are endless and include such things as the scrapings from shoes, bird droppings, road sweepings and the banks of drainage ditches. Some results of a simple experiment of this kind are shown in Fig. 6–4.

In all such experiments it is useful to take the sample soil and to mix it thoroughly with some form of sterile material such as John Innes compost or even coarse sand before placing it in pots or shallow trays. This has the virtues of making the growing conditions for different samples more nearly comparable and also minimizing difficulties due to the drainage properties of field soils, especially the heavier ones. In longer term experiments to look at things like the periodicity of germination it is also important to stir the top layers of the soil well every few weeks to make sure that a fresh supply of seeds is brought to the surface.

A particularly fascinating study of the light requirements of seeds in soil under pastures was made by WESSON and WAREING (1967) who showed that a very high percentage of the species represented (20 out of 23) had seeds which did not germinate over a period of twelve months unless exposed to light. The simplicity of the experimental methods used and the extremely spectacular differences between treatments which appeared illustrate the very interesting results which can be obtained in this kind of work.

6.3   Experiments on competition between plants

Even a cursory reading of chapter 2 will have indicated that this is an area of work fraught with difficulties. Nevertheless, it is one in which simple experiments can still give useful and meaningful results. Experiments can be conducted in pots, boxes of soil or in garden or field plots, and infinite variations on the theme are possible. A reading of one or two of the papers cited in chapter 2 will give more idea of the possibilities than can be conveyed in a small space here. Points particularly worth bearing in mind are that relative dates of sowing of a crop and weed(s) may be absolutely critical in determining the kind of result obtained; that different combinations of weeds, or different crop with the same weed, may give very different results; that the type of weed/crop combination chosen will greatly affect the length of time needed for the experiment and, finally, that particularly careful thought will be needed in relation to what is to be measured to achieve a meaningful result.

Fig. 6–4 Photographs of the result four weeks after four different soil samples (about 200–300 g each) were mixed into the surface layers of boxes of sterile John Innes potting compost. Two samples were used per box, one at each end as follows. (a) Sweepings from beside drain grating on road. (b) Arable field containing a cereal crop (c) Soil from author's gardening shoes. Note tomato plant, presumably from seed near compost heap. (d) Soil from Wye College experimental plot. Intensively cultivated for several years.

In conclusion all one can say is that a very little experience in experimenting with weeds will quickly show some of the problems which weed biologists all over the world are up against. Perhaps the experience may even stimulate you to want to join them in the study of this remarkable and fascinating group of plants.

References

ANDERSON, E. (1954). *Plants, Man and Life*. Andrew Melrose, London.

BAKER, H. G. (1965). In: Baker and Stebbins (1965) below.

BAKER, H. G. and STEBBINS, G. L. (1965). *The Genetics of Colonising Species*. Academic Press, New York.

BAKKER, D. (1960). In: Harper (1960) below.

BANDEEN, J. D. and BUCHOLTZ, K. B. (1967). *Weeds*, **15**, 220–4.

BLEASDALE, J. K. A. (1960). In: Harper (1960) below.

BUNTING, A. H. (1960). In: Harper (1960) below.

CHANCELLOR, R. J. (1966). *The Identification of Weed Seedlings of Farm and Garden*. Blackwell, Oxford.

CHANCELLOR, R. J. (1974). *Weed Res.*, **14**, 29–38.

CLAPHAM, A. R., TUTIN, T. G. and WARBURG, E. F. (1968). *Excursion Flora of the British Isles*. 2nd ed. Cambridge University Press, Cambridge.

CRAFTS, A. S. and ROBBINS, W. W. (1962). *Weed Control*. McGraw-Hill, New York.

CRAMER, H. H. (1967). *Plant Protection and World Crop Production*. (Translated by J. H. Edwards.) Bayer, Leverkusen.

CUMMING, B. G. (1963). *Can. J. Bot.*, **41**, 1211–33.

DAVISON, J. G. (1970). *Proc. 10th Br. Weed Control Conf.*, 352–7.

DAWSON, J. H. (1964). *Weeds*, **12**, 206–8.

DONY, J. G., ROB, C. M. and PERRING, F. H. (1974). *English Names of Wild Flowers*. Butterworth, London.

EHRENDORFER, F. (1965). In: Baker and Stebbins (1965) above.

FOGG, J. M. (1975). *Brooklyn bot. Gdn. Rec.*, **31**, 12–15.

FRYER, J. D. and EVANS, S. A. Eds (1968): *Weed Control Handbook*, Vol. I. *Principles*. 5th edn. Reprinted 1970. Blackwell, Oxford.

GODWIN, H. (1956). *The History of the British Flora*. Cambridge University Press, Cambridge.

GRÜMMER, G. and BAYER, H. (1960). In: Harper (1960) below.

HAIZEL, K. A. and HARPER, J. L. (1973). *J. appl. Ecol.*, **10**, 23–31.

HANF, M. (1972). *Weeds and their Seedlings*. B.A.S.F. Ipswich.

HARPER, J. L. Ed. (1960). *The Biology of Weeds*. British Ecological Society Symposium No. 1. Blackwell, Oxford.

HARPER, J. L. (1961). In: Milthorpe (1961) below.

HARPER, J. L. (1965). In: Baker and Stebbins (1965) above.

HARPER, J. L. and McNAUGHTON, I. H. (1960). *Heredity*, **15**, 315–20.

HARPER, J. L., WILLIAMS, J. T. and SAGAR, G. R. (1965). *J. Ecol.*, **53**, 273–86.

HEATH, O. V. S. (1970). *Investigation by Experiment*. Studies in Biology No. 23. Edward Arnold, London.

HITCHINGS, S. (1960). In: Harper (1960) above.

KEBLE MARTIN, W. (1965). *The Concise British Flora in Colour*. Ebury Press and Michael Joseph, London.

MARTIN, P. and RADEMACHER, B. (1960). In: Harper (1960) above.

MILTHORPE, F. L. Ed. (1961). *Mechanisms of Biological Competition*. Symp. Soc. exp. Biol. No. 15. Cambridge University Press, Cambridge.

MULLIGAN, G. A. (1965). In: Baker and Stebbins (1965) above.

MUZIK, T. J., (1970). *Weed Biology and Control*. McGraw-Hill, New York.

NYAHOZA, F., MARSHALL, C. and SAGAR, G. R. (1973). *Weed Res.*, **13**, 304–9.

PARKER, R. E. (1973). *Introductory Statistics for Biology*. Studies in Biology No. 43. Edward Arnold, London.

PAVLYCHENKO, T. K. and HARRINGTON, J. B. (1934). *Can. J. Res.*, **10**, 77–94.

PENNINGTON, W. (1974). *The History of British Vegetation*. 2nd edn. English Universities Press, London.

PRITCHARD, T. (1960). In: Harper (1960) above.

ROBERTS, H. A. (1958). *J. Ecol.*, **46**, 759–68.

ROBERTS, H. A. (1970). *Rep. natn. Veg. Res. Stn. for 1969*, 25–38.

ROBERTS, H. A., CHANCELLOR, R. J. and THURSTON, J. M. (1968). In: Fryer and Evans (1968) above.

SAGAR, G. R. (1968). *Neth. J. agric. Sci.*, **16**, 155–64.

SAGAR, G. R. and HARPER, J. L. (1960). In: Harper (1960) above.

SAKAI, K.-I. (1961). In: Milthorpe, F. L. (1961) above.

SALISBURY, SIR YE (1961). *Weeds and Aliens*. Collins, London.

SHADBOLT, C. A. and HOLM, L. G. (1956). *Weeds*, **4**, 111–23.

STEBBINS, G. L. (1971). *Chromosomal Evolution in Higher Plants*. Edward Arnold, London.

THURSTON, J. M., (1960). In: Harper (1960) above.

THURSTON, J. M. (1972). *Proc. 11th Br. Weed Control Conf.*, 977–87.

TINKER, J. (1974). *New Scient.*, **61**, 747–9.

TOOLE, E. J. and BROWN, E. (1946). *J. agric. Res.*, **72**, 201–10.

WELBANK, P. J. (1960). In: Harper (1960) above.

WELBANK, P. J. (1963). *Weed Res.*, **3**, 205–14.

WELLINGTON, P. S. (1960). In: Harper (1960) above.

WESSON, G. and WAREING, P. F. (1967). *Nature (Lond.)*, **213**, 600–1.

Appendix

List of plants mentioned in the text, with Latin and common names, families and whether annual (A), biennial (B) or perennial (P).

Latin name	Common name	Family	Habit
Achillea millefolium L.	Yarrow	Compositae	P
Aegopodium podagraria L.	Ground-elder	Umbelliferae	P
Ageratum conyzoides L.	—	Compositae	A
Ageratum microcarpum (Benth.) Hemsl.	—	Compositae	P
Agropyron repens (L.) Beauv.	Common Couch	Gramineae	P
Agrostemma githago L.	Corncockle	Caryophyllaceae	A
Agrostis gigantea Roth.	Black Bent	Gramineae	P
Alhagi camelorum Firsch.	Camel's Thorn	Leguminosae	P
Alopecurus myosuroides Huds.	Black-grass	Gramineae	A
Allium vineale L.	Wild Onion	Liliaceae	P
Anthriscus sylvestris (L.) Hoffm.	Cow Parsley	Umbelliferae	B
Armoracia rusticana Gaertn., Mey. & Scherb.	Horse-radish	Cruciferae	P
Avena fatua L.	Wild Oat	Gramineae	A
Avena ludoviciana Durieu.	Winter Wild Oat	Gramineae	A
Bellis perennis L.	Daisy	Compositae	P
Calluna vulgaris (L.) Hull	Heather	Ericaceae	P
Calystigia sepium (L.) R.Br.	Hedge or Large Bindweed	Convolvulaceae	P
Camelina alyssum (Mill.) Thellung.	—	Cruciferae	A
Camelina sativa (L.) Cranz.	Gold-of-pleasure	Cruciferae	A/B
Capsella bursa-pastoris (L.) Medic.	Shepherd's Purse	Cruciferae	A
Cardaria draba (L.) Descr.	Hoary Cress	Cruciferae	A/P
Chamenaerion Angustifolium (L.) Scop.	Rosebay Willow-herb	Onagraceae	P
Chenopodium album L.	Fat-hen	Chenopodiaceae	A
Chrysanthemum segetum L.	Corn Marigold	Compositae	A
Cirsium arvense (L.) Scop.	Creeping Thistle	Compositae	P
Convolvulus arvensis L.	Field Bindweed	Convolvulaceae	P
Conyza canadensis (L.) Cronq.	Canadian Fleabane	Compositae	A
Cyperus esculentus L.	Yellow Nutsedge	Cyperaceae	P
Cyperus rotundus L.	Purple Nutsedge	Cyperaceae	P
Draba muralis L.	Wall Whitlowgrass	Cruciferae	A/B

Latin name	Common name	Family	Habit
Eichornia crassipes Solms.	Water Hyacinth	Pontederiaceae	P
Eupatorium microstemon Cass.	—	Compositae	A
Eupatorium pycnocephalum Less.	—	Compositae	P
Fumaria officinalis L.	Common Fumitory	Fumariaceae	A
Galium aparine L.	Cleavers	Rubiaceae	A
Geranium dissectum L.	Cut-leaved Cranesbill	Geraniaceae	A
Helianthus annuus L.	Sunflower	Compositae	A/P
Hypericum perforatum L.	Perforate St John's Wort	Hypericaceae	P
Juncus inflexus L.	Hard Rush	Juncaceae	P
Lamium purpureum L.	Red Dead-nettle	Labiatae	A
Leontodon spp.	Hawkbits	Compositae	P
Lychnis spp.	Ragged Robin. Catchfly	Caryophyllaceae	P
Matricaria matricarioides (Less.) Porter	Pineappleweed	Compositae	A
Opuntia spp.	Prickly Pear	Cactaceae	P
Oxalis pes-caprae L.	Bermuda Buttercup	Oxalidaceae	P
Papaver rhoeas L.	Common Poppy	Papaveraceae	A
Petroselinum segetum (L.) Koch.	Corn Parsley	Umbelliferae	B
Plantago lanceolata L.	Ribwort Plantain	Plantaginaceae	P
Plantago major L.	Greater Plantain	Plantaginaceae	P
Plantago media L.	Hoary Plantain	Plantaginaceae	P
Poa annua L.	Annual Meadowgrass	Gramineae	A
Polygonum aviculare L.	Knotgrass	Polygonaceae	A
Polygonum convolvulus L.	Black Bindweed	Polygonaceae	A
Polygonum persicaria L.	Redshank	Polygonaceae	A
Pteridium aquilinum (L.) Kühn.	Bracken	Polypodiaceae	P
Ranunculus repens L.	Creeping Buttercup	Ranunculaceae	P
Raphanus raphanistrum L.	Wild Radish	Cruciferae	A
Rumex acetosella L.	Sheep's Sorrel	Polygonaceae	P
Rumex crispus L.	Curled Dock	Polygonaceae	P
Rumex obtusifolius L.	Broad-leaved Dock	Polygonaceae	P
Sagina procumbens L.	Procumbent Pearlwort	Caryophyllaceae	P
Salvina spp.	—	Salviniaceae	
Senecio jacobaea L.	Common Ragwort	Compositae	B/P*
Senecio squalidus L.	Oxford Ragwort	Compositae	A/B*/P*

Latin name	Common name	Family	Habit
Senecio vulgaris L.	Groundsel	Compositae	A
Sinapis alba L.	White Mustard	Cruciferae	A
Sinapis arvensis L.	Charlock	Cruciferae	A
Solanum nigrum L.	Black Nightshade	Solanaceae	A
Solidago canadensis L.	Canadian Golden-rod	Compositae	P
Sonchus asper (L.) Hill	Prickly Sowthistle	Compositae	A/B
Sorghum halepense (L.) Pers.	Johnson Grass	Gramineae	P
Stellaria media (L.) Vill.	Common Chick-weed	Caryophyllaceae	A
Striga spp.	Witchweeds	Scrophulariaceae	A
Taraxacum officinale Weber	Dandelion	Compositae	P
Tridax repens L.	—	Compositae	P
Tussilago farfara L.	Colt's-foot	Compositae	A
Urtica urens L.	Small Nettle	Urticaceae	A

* Rarely